ESPRIT Basic Research Series

Edited in cooperation with
the Commission of the European Communities, DG XIII

J. W. Lloyd (Ed.)

Computational Logic

Symposium Proceedings
Brussels, November 13/14, 1990

Springer-Verlag Berlin Heidelberg New York
London Paris Tokyo Hong Kong Barcelona

Volume Editor

John W. Lloyd
University of Bristol
Department of Computer Science
Queen's Building, University Walk
Bristol BS8 1TR, UK

ISBN 3-540-53437-7 Springer-Verlag Berlin Heidelberg New York
ISBN 0-387-53437-7 Springer-Verlag New York Berlin Heidelberg

Publication No. EUR 13147 EN of the
Commission of the European Communities,
Scientific and Technical Communication Unit,
Directorate-General Telecommunications, Information Industries and Innovation,
Luxembourg
Neither the Commission of the European Communities nor any person acting on behalf of the
Commission is responsible for the use which might be made of the following information.

Printing: Druckhaus Beltz, Hemsbach; Binding: J. Schäffer GmbH & Co. KG, Grünstadt
2145/3140-543210 – Printed on acid-free paper

FOREWORD

This volume has a dual significance to the ESPRIT Basic Research efforts towards forging strong links between European academic and industrial teams carrying out research, often interdisciplinary, at the forefront of Information Technology.

Firstly, it consists of the proceedings of the "Symposium on Computational Logic" – held on the occasion of the 7th ESPRIT Conference Week in November 1990 – whose organisation was inspired by the work of Basic Research Action 3012 (COMPULOG). This is a consortium which has attracted world-wide interest, with requests for collaboration throughout Europe, the US and Japan. The work of COMPULOG acts as a focal point in this symposium which is broadened to cover the work of other eminent researchers in the field, thus providing a review of the state of the art in computational logic, new and important contributions in the field, but also a vision of the future.

Secondly, this volume is the first of an ESPRIT Basic Research Series of publications of research results. It is expected that the quality of content and broad distribution of this series will have a major impact in making the advances achieved accessible to the world of academic and industrial research alike.

At this time, all ESPRIT Basic Research Actions have completed their first year and it is most encouraging and stimulating to see the flow of results such as the fine examples presented in this symposium.

This is a good beginning; and there is much more to come!

George Metakides

Contributors

K.R. Apt

A. Bundy

A. Colmerauer

R.A. Kowalski

V. Lifschitz

J. McCarthy

R. Reiter

D.S. Scott

J.H. Siekmann

F. Turini

Panellists

G. Comyn

H. Gallaire

R.A. Kowalski

J. McCarthy

D.S. Scott

Programme Committee

H. Gallaire

R.A. Kowalski

G. Levi

J.W. Lloyd

PREFACE

This Symposium brings together 10 leading researchers in computational logic, each of whom was invited to contribute a paper. The contributors are Krzysztof Apt, Alan Bundy, Alain Colmerauer, Robert Kowalski, Vladimir Lifschitz, John McCarthy, Raymond Reiter, Dana Scott, Jörg Siekmann, and Franco Turini.

The history of computational logic can be traced back at least to Hobbes and Leibniz in the 17th century, Boole and Frege in the 19th century, Herbrand in the 1930's, through to more recent work of J.A. Robinson in the 1960's. The basic idea of computational logic is that the model-theoretic semantics of logic provides a basis for knowledge representation and the proof-theoretic semantics provides a basis for computation. Computational logic is not a single field of Computer Science, but encompasses fields such as logic programming, functional programming, database systems, formal methods, and substantial parts of artificial intelligence.

The importance of computational logic is growing rapidly. This can be seen in the proliferation of textbooks on logic for Computer Science. A course on logic is now widely seen to be an essential requirement in an undergraduate Computer Science course. Its importance can also be seen in the rapid growth of the use of logic and functional languages, formal methods, and logic in database systems and artificial intelligence, and in the increasing exploitation of parallelism inherent in logical formalisms.

The contributed papers range from some on specialised research topics to some which give an overview and a glimpse into the future of computational logic.

Krzysztof Apt and his colleague provide in their paper a theoretical basis for studying termination of logic programs using the Prolog selection rule. They introduce the class of left terminating programs, which are those that terminate with the Prolog selection rule for all ground goals. They prove a characterization of left terminating programs which provides a practical method for proving termination. They illustrate their method by giving simple proofs of termination of several well-known programs.

In their paper, Alan Bundy and his colleagues show how to synthesise logic programs from non-executable specifications. The technique is adapted from one for synthesising functional programs as total functions. The key idea of the adaptation is that a predicate can be regarded as a total function when all its arguments are ground. The program is synthesised as a function in this mode and then run in other modes. The technique has been tested on the OYSTER program development system.

Alain Colmerauer introduces the language Prolog III in his paper. Prolog III extends Prolog by replacing unification by constraint solving. It integrates processing of trees and lists, numerical processing, and Boolean algebra. Colmerauer presents the specification and theoretical foundations of the new language. In fact, the theoretical foundations apply to a whole family of languages. He also demonstrates the capabilities of Prolog III with a variety of applications.

In his paper, Robert Kowalski is concerned with the development of the use of logic for all aspects of computation. This includes not only the development of a single logic for representing programs, program specifications, databases, and knowledge representations in artificial intelligence, but also the development of logic-based management tools. He argues that, for these purposes, two major extensions of logic programming are needed – abduction and metalevel reasoning.

Vladimir Lifschitz proposes in his paper a new form of default logic in which the parameters in a default are treated as genuine object variables, instead of the usual treatment as meta-variables for ground terms. This new form of default may be preferable when the domain closure assumption is not made. Lifschitz shows that this form of default has a particularly simple relation to circumscription.

The paper by John McCarthy is concerned with two kinds of program specifications, which he calls *illocutionary* and *perlocutionary*. The first are input-output specifications and the second accomplishment specifications. McCarthy discusses the difference between the two. He also discusses how the use of formal methods in proving perlocutionary specifications requires formalization of assumptions about the world.

The paper by Raymond Reiter studies the application of an epistemic modal logic to databases. In this approach, a database is a first order theory, but queries and constraints are expressed in the more expressive epistemic modal logic KFOPCE, introduced by Levesque. Reiter shows how to do query evaluation and integrity maintenance, and provides a sufficient condition for the completeness of the query evaluator.

The paper by Dana Scott is concerned with the symbolic computation program *Mathematica*. He discusses the use of this program in two courses, one on projective geometry and one on methods of symbolic computation. He then discusses several lessons gained from the experience and gives suggestions for future developments.

Jörg Siekmann and his colleagues address in their paper the combination of fast taxonomical reasoning algorithms that come with concept languages used as knowledge representation formalisms and reasoning in first order predicate logic. The interface between these two different modes of reasoning is accomplished by a new rule of inference, called constrained resolution. They show the correctness, completeness, as well as the decidability of the constraints (in a restricted constraint language).

The paper by Franco Turini and his colleagues is concerned with an algebra of operators for logic programs. The operators considered are analogous to set-theoretic union, intersection, and difference. They provide both a transformational and an interpretive characterization of the operators and prove them equivalent. They also present some examples from default reasoning, knowledge assimilation, inheritance networks, and hypothetical reasoning to demonstrate the expressive power of the operators.

A major theme of the Symposium is the likely future impact of computational logic and this is emphasised by the panel session on the topic "Programming in 2010: The Role of Computational Logic". The panellists are Gérard Comyn, Hervé Gallaire, Robert Kowalski, John McCarthy, and Dana Scott. The position papers of the panellists are contained in these proceedings.

The task presented to the panellists was to consider the major issues in programming and to try to predict what programming will be like 20 years from now, with particular emphasis on the role that computational logic may play. Thus, the panellists were asked to address questions such as:

- Will program synthesis be practical in 2010?

- Will automatic program verification be practical in 2010?

- How will logic and functional programming have evolved in the next 20 years?

- What role will logic play in artificial intelligence in 2010?

- What contribution will logic make to programming environments in the next 20 years?

- What kinds of computer architecture will support computational logic in 2010?

It will be very interesting to compare in 20 years the predictions made by the panellists to what has actually happened then!

This Symposium could not have taken place without the considerable efforts of a number of people. First, George Metakides had the original inspiration for the Symposium and made the resources of ESPRIT Basic Research Actions available. The other programme committee members were Giorgio Levi, Hervé Gallaire, and Robert Kowalski. The programme committee members chose the contributors. They also advised them on their papers. Particular thanks are due to the contributors, who are to be congratulated for producing papers of such high quality in such a short time and for their excellent co-operation. Finally, thanks are due to Ingo Hussla of EBRA for a considerable amount of behind-the-scenes organization.

J.W. Lloyd

CONTENTS

PROBLEMS AND PROMISES OF COMPUTATIONAL LOGIC

Robert A. Kowalski

Imperial College, Department of Computing,

180 Queen's Gate, London SW7 2BZ, UK

Abstract

The ultimate goal of the Basic Research Action, Compulog, is to develop the use of logic for all aspects of computation. This includes not only the development of a single logic for representing programs, program specifications, databases, and knowledge representations in artificial intelligence, but also the development of logic-based management tools. I shall argue that, for these purposes, two major extensions of logic programming are needed - abduction and metalevel reasoning. I shall also argue that extensions of full first-order logic may not be necessary.

0.   Introduction

The term "computational logic" has no generally agreed meaning. Here I shall use the term, in the sense in which it is used in the ESPRIT Basic Research Action "Compulog", as the use of logic for all aspects of computing: not only for representing programs, program specifications, databases, and knowledge bases; but also for processing, developing, and maintaining them.

Historically, the techniques of computational logic have arisen from work on the automation of logical deduction, begun by logicians in the 1950s. This resulted during the 1970s in the development of efficient theorem-proving techniques, based on the resolution principle, for processing logic programs. More recent work has focussed on developing techniques for deductive databases and for default reasoning in artificial intelligence. Within Compulog we aim to develop unified logic-based techniques for knowledge representation, knowledge processing, and knowledge assimilation applicable to the three areas of programming, databases and artificial intelligence.

In this paper I shall present a personal view of some of the characteristic achievements and problems of computational logic. I shall touch upon some, but not all, of the work areas of Compulog. I will focus in particular upon two major topics in Compulog: knowledge assimilation and metareasoning. I will not deal with three other Compulog topics: constraint logic programming, structured types, and program development, transformation and analysis. These other topics are addressed in other papers presented at the Symposium. I will also discuss the broader question of what subset of full first-order logic is needed for practical applications.

1. Knowledge processing, representation, and assimilation

Logic has traditionally focussed upon the problem of determining whether a given conclusion (or theorem) C is logically implied by a given set of assumptions (or theory) T. Computational logic in particular has been concerned with developing efficient theorem-proving methods for determining such implications. It has also been concerned with applying such theorem-provers as "knowledge processors", for theories representing logic programs, deductive databases, or knowledge bases, and for theorems representing procedure calls, database queries, or problems to be solved.

In computing we need to be concerned not only with knowledge processing, but also with knowledge representation and knowledge assimilation. Knowledge representation concerns the way in which knowledge is formalised. Because the same knowledge can generally be formalised in many different ways, the subject of knowledge representation is concerned with identifying useful guidelines for representation. In programming, for example, these guidelines deal with such matters as the choice of data structures, recursive versus iterative procedures, and the treatment or avoidance of side effects; in databases, with normal forms, nested relations and metadata; and in artificial intelligence, with such issues as the treatment of defaults and uncertainty.

Knowledge assimilation concerns the assimilation of new knowledge into an existing theory. The simplest case is that of an update to be added to a database. The update is rejected if it fails to satisfy certain integrity constraints. Otherwise it is inserted into the database. More elaborate cases of knowledge assimilation can occur if the theory has richer logical structure. Four cases stand out:

i) The update is the addition of a sentence which can be deduced from the existing theory. In this case the update can be ignored, and the new theory can remain the same as the old theory.

ii) The update is the addition of a sentence which together with some sentences in the existing theory can be used to derive some of the others. In this case the update can replace the derivable old sentences.

iii)   The update can be shown to be incompatible with the existing theory and integrity constraints. In this case a belief revision process needs to determine how to restore satisfaction of the integrity constraints. This might non-deterministically involve rejection or modification of any of the assumptions which partake in the proof of incompatibility.

iv)   The update is logically independent from the existing theory, in the sense that none of the above relationships (i), (ii) or (iii) can be shown. In this case the simplest way to construct the new theory is to insert the update into the old theory. In many situations, however, it may be more appropriate to generate an abductive explanation of the update and to insert the explanation in place of the update into the old theory. By the definition of abduction, the explanation is so constructed that the original update can be derived from the new theory and the new theory satisfies any integrity constraints. The derivation of such an abductive explanation might be non-deterministic, and result in several alternative new theories.

Thus knowledge assimilation is a significant extension of database updates. It can play a major role in the incremental development and maintenance of logic-based programs, program specifications, and knowledge bases.

The topics of knowledge processing, representation, and assimilation are interrelated, and will be addressed in the remaining three sections of this paper. Section (2), which deals with abduction and integrity constraints, is specifically concerned with knowledge assimilation, but is also concerned with procedures for performing abduction and integrity checking, as well as with representing defaults. Section (3) deals with metareasoning, which is an important technique for specifying and implementing both theorem-provers and knowledge assimilators. It also considers the use of metalogic for representing modalities such as knowledge and belief. Section (4) deals with the question of whether full first-order logic is necessary for knowledge representation, or whether extensions of logic programming form might be more useful. Although essentially a knowledge representation matter, this issue has important consequences for the kinds of theorem-provers and knowledge assimilators needed for practical applications.

2.   Abduction and integrity constraints

Abduction seems to play an important role in everyday human problem solving. Charniak and McDermott, for example, in their "Introduction to Artificial Intelligence" [51] argue that abduction plays a key role in natural language understanding, image recognition and fault diagnosis; Poole [42] argues that it can be used to give a classical account of default reasoning; Eshghi and Kowalski [14] argue that it generalises negation by failure in logic programming; and within Compulog, Bry [7] and Kakas and Mancarella [24] have shown that abduction can be used to solve the view update problem in deductive databases. Several authors [12, 14, 16] have shown that it can be implemented naturally and efficiently by a simple modification of backward reasoning.

Given a theory T, integrity constraints I, and a possible conclusion C, an abductive explanation of C is a set of sentences Δ such that

> $T \cup \Delta$ logically implies C, and
> $T \cup \Delta$ satisfies I.

The notion of integrity constraint arises in the field of databases as a property which a database is expected to satisfy as it changes over the course of time. As Reiter [44] explains in his contribution to this Symposium, different ways to formalise constraints and to define constraint satisfaction have been proposed. For present purposes, and for simplicity's sake, I shall assume that constraints are formulated as denials, and that constraint satisfaction is interpreted as consistency in a logic with an appropriate semantics for negation by failure. Such a semantics might be given, for example, either by the completion of the theory [9] or by stable models [17].

I shall also assume that the theory is represented as a set of clauses in logic programming form

> $A_0 \leftarrow A_1, ..., A_n$

where the conclusion A_0 is an atomic formula and the conditions A_i, $1 \leq i \leq n$, are atomic formulae or negations of atomic formulae. Constraints are represented as denials of the form

> $\leftarrow A_1, ..., A_n$

Problems to be solved also have the form of denials, but are more naturally expressed with the negation symbol written as a question mark.

> $? A_1, ..., A_n$

2.1 Fault diagnosis

The following simplified example of some of the causes of bicycle faults shows that abduction is a natural procedure for performing fault diagnosis.

Theory: wobbly-wheel ← flat-tyre

                        wobbly-wheel ← broken-spokes

                        flat tyre ← punctured-tube

                        flat tyre ← leaky-valve

Constraint: ← flat-tyre, tyre-holds-air

To determine the possible causes of wobbly-wheel, it suffices to seek the abductive explanations of wobbly-wheel. This can be performed simply by reasoning backward, logic programming style, from the goal

> ? wobbly-wheel.

Instead of failing on the subgoals

> ? punctured-tube
> ? leaky-valve
> ? broken-spokes

these subgoals can be assumed to hold as hypotheses, provided they are consistent with the theory and integrity constraint, which they are.

In this example the three subgoals correspond to three alternative hypotheses, which give rise in turn to three alternative new theories, each of which explains the conclusion. Given the lack of further information such multiple solutions are unavoidable.

Given the additional information

> tyre-holds-air

however, the two hypotheses

> punctured-tube
> leaky-valve

become inconsistent and have to be withdrawn non-monotonically.

Fault diagnosis can also be performed by deduction alone, as it is in most current expert systems. However, this requires that the natural way of expressing laws of cause and effect be reversed. This can be done in two quite different ways, one using logic programming form, the other using non-Horn clauses.

In the bicycle faults example, ignoring the integrity constraint, the deductive, logic programming representation might take the form

    possible(flat-tyre) ← possible(wobbly-wheel)
    possible(broken-spokes) ← possible(wobbly-wheel)
    possible(punctured-tube) ← possible(flat-tyre)
    possible(leaky-valve) ← possible(flat-tyre)
    possible(wobbly-wheel)
 ? possible(X)

Even ignoring the problem of representing the integrity constraint, compared with the abductive formulation, this formulation is lower-level - more like a program than like a program specification.

The alternative, non-Horn representation has the form

    flat-tyre v broken-spokes ← wobbly-wheel
    punctured-tube v leaky-valve ← flat-tyre
    wobbly-wheel

In this formulation it is natural to reason forward rather than backward, deriving the disjunction

    punctured-tube v leaky-valve v broken-spokes

of all possible causes of the fault. This formulation has some advantages, including the ease with which it is possible to represent integrity constraints. Given, for example, the additional sentences

    ← flat-tyre, tyre-holds-air
    tyre-holds-air

it is possible monotonically to derive the new conclusion

broken-spokes.

The non-Horn formulation of the bicycle faults problem is an example of a general approach to abduction through deduction proposed by Console et al [10].

It also illustrates the phenomena, to be discussed in section (4) below, that many problems can be represented both in logic programming and non-Horn clause form. (In this case, it is easy to see that the non-Horn formulation is the only-if half of the completion of the Horn clause formulation.) My own intuition is that logic programming form is simpler and easier to implement, but that it entails the need for extensions such as abduction and integrity constraints. It may also be, however, that the non-Horn clause form is useful for certain purposes.

2.2   Default reasoning

Poole [42] was perhaps the first to argue that abduction provides a natural mechanism for performing default reasoning.   This can be illustrated with Reiter's example [43] of conflicting defaults. Expressed as a logic program with abduction and integrity constraints this can be represented in the form

Theory: pacifist(X) ← quaker(X), normal-quaker(X)
              hawk(X) ← republican(X), normal-republican(X)
              quaker(nixon)
              republican(nixon)

Constraint: ← pacifist(X), hawk(X)

It is possible to explain both the conclusion

pacifist(nixon)   with the hypothesis
normal-quaker(nixon)

and the conclusion

hawk(nixon)   with the hypothesis
normal-republican(nixon).

Each hypothesis on its own is consistent with the integrity constraint, but together the hypotheses are inconsistent. This phenomenon of conflicting defaults is sometimes thought to be a problem with default reasoning, but it is a natural characteristic of abduction. Moreover, as Poole [42] has shown, maximally consistent sets of abductive hypotheses correspond to extensions in Reiter's default logic.

The problem of accepting the existence of conflicting defaults seems to be partly psychological and partly technical. If default reasoning is viewed as deduction, then it is natural to view extensions in default logic as models, and to regard logical consequence as truth in all models. But proof procedures for default logic determine only what holds in a single extension, rather than what holds in all extensions. However, if default reasoning is viewed as abduction, then it is natural to view extensions as theories, and to view multiple extensions as alternative theories. In this case, proof procedures for determining what holds in single extensions are exactly what is needed.

The number of alternative abductive hypotheses can often be reduced by assigning priorities between them. Such priorities can be formulated as statements in the theory itself, as shown by Pereira and Aparicio [40].

2.3 Negation by failure

Negation by failure is a simple and efficient form of default reasoning in logic programming. It has been shown [14] that abduction simulates and generalises negation by failure. This can be illustrated most simply with a propositional example.

> p ← not q
> q ← r

The negative condition can be thought of as a possible abductive hypothesis:

> not q may be assumed to hold
> if it is consistent to assume it holds.

That it is inconsistent to hold both an atom and its negation can be expressed by the integrity constraint

> ← q, not q

It is now possible to show p by means of backward reasoning with abduction and integrity checking, where not q is treated as a composite positive atom.

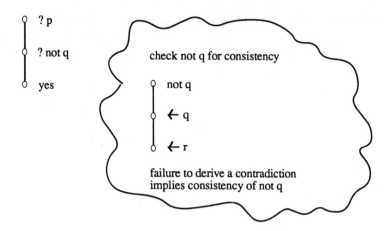

Here the possible hypothesis not q is tested by forward reasoning, as in the consistency method of integrity checking [45]. It is easy to see in this case that abduction with integrity checking simulates negation by failure.

The situation is more complex with nested negation in examples like

$p \leftarrow$ not q
$q \leftarrow$ not r

With only the two integrity constraints

$\leftarrow$ q, not q
$\leftarrow$ r, not r

The test for consistency of not q has two possible outcomes:

1)   not q is inconsistent
      with the hypothesis not r.
      So p fails.

2)   not q is consistent
      without the hypothesis not r.
      So p succeeds.

The first outcome accords with negation by failure, the second does not.

The second outcome can be eliminated by adding an extra integrity constraint

r v not r

which is viewed as a metalevel or modal epistemic statement [44] that

either r or not r
can be derived from the theory.

This eliminates the second outcome, because, since r cannot be proved, the hypothesis not r must be assumed. The atom r cannot be proved because, when abduction is used simply to simulate negation by failure, only negative atoms are assumed as hypotheses.

Under the interpretation of negative conditions as assumable hypotheses, abduction with appropriate integrity constraints simulates negation by failure. Moreover, there is a simple one-to-one correspondence between sets of abducible hypotheses satisfying the integrity constraints and stable models [14] of the original logic program.

Under the same interpretation, abduction with integrity checking generalises negation by failure. For a program such as

p ← not q
q ← not p

it gives the two alternative outcomes

1)   p with the hypothesis not q
2)   q with the hypothesis not p,

corresponding to the two stable models of the program. As in fault diagnosis and other applications of abduction, such multiple alternatives are quite acceptable.

2.4   The Yale Shooting Problem

This problem was identified by Hanks and McDermott [19] as an example of default reasoning where circumscription [36] and default logic [43] give intuitively unacceptable results. It is interesting to note that the example has the form of a logic program, and that negation by failure gives the correct result.

Here

   t(P, S)

expresses that property P holds in situation S,

   ab(P, E, S)

expresses that P is abnormal with respect to event E occurring in situation S, and the term

   result(E, S)

names the situation that results from the occurrence of event E in situation S.

   t(alive, s0)
   t(loaded, result(load, S))
   t(dead, result(shoot, S)) ← t(loaded, S)
   t(P, result(E, S)) ← t(P, S), not ab(P, E, S)
   ab(alive, shoot, S) ← t(loaded, S)

In circumscription, minimising the predicate "ab", the theory has two minimal models. In one model the atom

1)   t(dead, result(shoot, result(wait, result(load, s0))))

is true. In the other the atom

2)   t(alive, result(shoot, result(wait, result(load, s0))))

is true. Defining logical consequence as truth in all models means that using circumscription the disjunction of (1) and (2) is a consequence of the theory. In the corresponding formulation in default logic, (1) and (2) hold in alternative extensions.

Hanks and McDermott argue that the model in which (1) holds is not intuitively acceptable. In the body of literature which has since been devoted to the problem, few commentators have observed that negation by failure gives the intuitively correct result [2, 14, 15].

Under the interpretation of negation by failure as a special case of abduction with integrity constraints, it can be argued that what is missing in the circumscriptive and default logic formulations of the problem is an appropriate analogue of the disjunctive integrity constraints [14]. Another abductive solution to the Yale shooting problem has been presented by [22].

2.5 Rules and exceptions

Closely related to abduction is another extension of logic programming in which clauses can have negative as well as positive conclusions. Clauses with positive conclusions represent general rules, and clauses with negative conclusions represent exceptions. This can be illustrated by the following formalisation of a familiar example.

Rules: fly(X) ← bird(X)
               bird(X) ← ostrich(X)
               bird(X) ← penguin(X)
               bird(tweety)
               ostrich(ozzy)

Exceptions: ¬ fly(X) ← ostrich(X)
               ¬ fly(X) ← penguin(X)

The stable model semantics of Gelfond and Lifschitz [17] can be extended and the answer set semantics [18] modified to give a natural semantics for such extended logic programs [32]. In this example, under this semantics, it is possible to conclude

      fly(tweety)
      ¬ fly(ozzy)

but not to conclude

      fly(ozzy).

The semantics of rules and exceptions is related to the semantics of abduction in that, like abducibles, general rules hold by default, unless they are contradicted by exceptions, which behave like integrity constraints. This relationship is easy to see in the following reformulation of the quaker-republican example:

Rules: pacifist(X) ← quaker(X)

                   hawk(X) ← republican(X)

                   quaker(nixon)

                   republican(nixon)

Exceptions: ¬ pacifist(X) ← hawk(X)

                 ¬ hawk(X) ← pacifist(X)

Under the extended semantics it is possible to conclude

                   pacifist(nixon) in one "model", and

                   hawk(nixon) in another.

The different "models", however, are more like different abductive theories than they are like ordinary models. In particular, it is of greater relevance to discover what those "models" are and what holds in them, than it is to determine what holds in all of them.

Exceptions are more specialised than integrity constraints, in that they also indicate how to restore consistency in case they conflict with general rules. For example, compared with the constraint

     ← fly(X), ostrich(X)

the exception

    ¬ fly(X) ← ostrich(X)

indicates that in the case of a conflict between conclusions of the form

       fly(t)

       ostrich(t)

it is the former conclusion rather than the latter that should be withdrawn.

The two exceptions in the second formulation of the quaker-republican example correspond to the one integrity constraint in the earlier abductive formulation. It is also possible in the rules and exceptions formulation to have only one exception rather than two. The resulting representation would then have only one "model" rather than two.

It is possible to transform rules and exceptions into normal logic programs with negation by failure [32]. This is most easily performed in two stages. In the first stage extra conditions of the form

not ¬ A

are introduced into rules having conclusions of the form A, whenever there are exceptions with conclusions of the form ¬ A. (Appropriate account must be taken of the case where the atoms of the conclusions are unifiable rather than identical). In the second stage negated atoms of the form ¬ A are replaced by positive atoms with some new predicate symbol. This is easy to illustrate with the two examples above.

In the case of the first example, after the first stage we obtain the clauses

fly(X) ← bird(X), not ¬ fly(X)
bird(X) ← ostrich(X)
bird(X) ← penguin(X)
bird(tweety)
ostrich(ozzy)
¬ fly(X) ← ostrich(X)
¬ fly(X) ← penguin(X)

Using a new predicate symbol "abnormal" in place of "¬fly", after the second stage we obtain the normal logic program

fly(X) ← bird(X), not abnormal(X)
bird(X) ← ostrich(X)
bird(X) ← penguin(X)
bird(tweety)
ostrich(ozzy)
abnormal(X) ← ostrich(X)
abnormal(X) ← penguin(X)

The transformation preserves the meaning of the original formulation in the sense that the two formulations have the same "models".

In the case of the second example, after the two stages we obtain the program

    pacifist(X) ← quaker(X), not ab1(X)
    hawk(X) ← republican(X), not ab2(X)
    quaker(nixon)
    republican(nixon)
    ab1(X) ← hawk(X)
    ab2(X) ← pacifist(X)

This can be further simplified by "macro-processing" (or partially evaluating) the new predicates.

    pacifist(X) ← quaker(X), not hawk(X)
    hawk(X) ← republican(X), not pacifist(X)
    quaker(nixon)
    republican(nixon)

The resulting program (after one further simplification step) has the familiar form

    p ← not q
    q ← not p

and has the same two "models" as the original formulation.

2.6 Semantic issues

The "model" theoretic semantics of rules and exceptions facilitates the proof that the transformation from rules and exceptions into normal logic programs preserves their meaning. It is also possible to generalise the stable model semantics of negation by failure to more general logic programs with abduction and integrity constraints [23]. It is an open problem whether such modifications of the semantics are really necessary.

As an alternative to modifying the semantics, it may be possible to view the meaning of these extensions of logic programming in metatheoretic terms. We have already remarked, in particular, that stable models behave in some respects more like theories than like conventional models. Moreover, our initial definition of abduction was formulated in metatheoretic rather than model theoretic terms.

Viewing the semantics of abduction and of rules and exceptions in terms of theories instead of models has the attraction that the underlying semantics remains that of the underlying deductive logic. This conservative approach to semantics has the advantage that seeming extensions of the logic are not extensions at all, but are defined in terms of metatheoretic relationships between object level theories. Just such a metatheoretic "semantics" for a special case of rules and exceptions can be found in the contribution by Brogi et al [6] presented at this Symposium. It has to be admitted, however, that extending this metatheoretic semantics to the general case does not seem to be entirely straight-forward.

A related problem concerns the semantics of integrity constraints. In his contribution to the Symposium, Reiter [44] argues that integrity constraints should be understood as statements about what the theory "knows". I agree with this interpretation, but propose that knowledge be understood metatheoretically in terms of what can be proved from the theory. This is a topic for the next section.

3. Metareasoning

A metalanguage is a language in which the objects of discussion are themselves linguistic entities. A metalogic, moreover, is a metalanguage which is also a logic.

Many applications of logic are inherently and unavoidably metalogical - applications, which formalise proof procedures (also called metainterpreters) and knowledge assimilators, for example. Metalogic can also be used for applications which can be formalised using other logics, such as modal logics of knowledge and belief. In this section I shall discuss some of these applications and their implementations.

3.1 The demo predicate

For many of these applications it is useful to employ a two-argument proof predicate

    demo(T, P)

which expresses that the theory named T can be used to demonstrate the conclusion named P. For other applications the first argument might be omitted or other arguments might be added - to name a proof of the conclusion or the number of steps in a proof, for example.

Perhaps the most common and most characteristic application of metalogic is to define and implement metainterpreters. Probably the simplest of these is a metainterpreter for propositional Horn clause programs.

(pr1) demo(T, P) ← demo(T, P ← Q),
$\qquad\qquad$ demo(T, Q)

(pr2) demo(T, P ∧ Q) ← demo(T, P),
$\qquad\qquad$ demo(T, Q)

Here "∧" names conjunction and (without danger of ambiguity) "←" names "←".

Theories can be named by lists (or sets) of names of their sentences, in which case we would need an additional metaclause such as

$\qquad$ demo(T, P) ← member(P, T)

together with a definition of the member predicate.

Execution of the metainterpreter, for given names of a theory and proposed conclusion, simulates execution of the object level theory and conclusion. For example, execution of the object-level program and query

$\qquad$ p ← q, r
$\qquad$ q ← s
$\qquad$ r
$\qquad$ s
$\qquad$? p

can be simulated by executing the metaquery

$\qquad$? demo([p ← q ∧ r, q ← s, r, s], p)

using the meta interpreter. Here, for simplicity's sake, I have not distinguished between atomic formulae and their names.

For some applications, such as the implementation of program transformation systems, for example, naming theories by lists is both appropriate and convenient. For other applications, however, it is cumbersome and inconvenient. In many simple cases, moreover, the theory argument can be eliminated altogether as in the so-called "vanilla" metainterpreter [20]. In the case of the simple object-level program given above, for example, execution of the object-level could be simulated by adding metalevel clauses to describe the object-level program.

    demo(p ← q ∧ r)
    demo(q ← s)
    demo(r)
    demo(s)

and posing the metalevel query

    ? demo(p)

using the same metainterpreter as before, but without the theory argument.

The vanilla metainterpreter uses different predicates for clauses which can be demonstrated because they are axioms (the predicate "clause") and clauses which can be demonstrated because they are derivable by means of one or more steps of inference (the predicate "solve"). I prefer the formulation presented above because of its greater generality and because the use of a single predicate emphasizes the similarity between the axioms of metalogic and the distribution axiom of modal logic.

Another curious feature of the vanilla metainterpreter is that it can be used for programs containing variables as well as for variable-free programs. This can be seen, for example, with the object-level program and query

    p(X) ← q(X)
    q(a)
    q(b)
    ? p(Y)

which can be represented (incorrectly!) at the metalevel by

    demo(p(X) ← q(X))
    demo(q(a))
    demo(q(b))
    ? demo(p(Y))

The representation is incorrect because the implicit quantifiers are treated incorrectly. The object-level clause is implicitly universally quantified

    ∀X[p(X) ← q(X)]

Instead of expressing that this universally quantified clause belongs to the program, the metalevel clause expresses

$\forall X demo(p(X) \leftarrow q(X))$

namely that for every term t the clause

$p(t) \leftarrow q(t)$

belongs to the program. It expresses in particular that the clauses

$p(a) \leftarrow q(a)$
$p(b) \leftarrow q(b)$

belong to the program.

This explanation of the meaning of the metalevel clause also explains why the vanilla metainterpreter works for programs containing variables, even though it is, strictly speaking, incorrect. The representation of object level variables by metalevel variables implicitly builds into the representation of the axioms and of the conclusion to be proved the correct rules

(pr3) $demo(Q) \leftarrow demo(forall(X, P))$,
$\quad\quad\quad\quad\quad substitute(X, P, Y, Q)$

(pr4) $demo(exists(X, P)) \leftarrow demo(Q)$,
$\quad\quad\quad\quad\quad substitute(X, P, Y, Q)$

where the predicate

$substitute(X, P, Y, Q)$

expresses that Q is the expression that results from substituting the term Y for the variable X in P. Given a correct metalevel representation, for example:

$demo(forall(var(1), p(var(1)) \leftarrow q(var(1))))$
$demo(q(a))$
$demo(q(b))$
$? demo(exists(var(2), p(var(2))))$

the (otherwise) incorrect metalevel representation can be derived by reasoning forward with the rule (pr3) and backward with the rule (pr4). This is because when the condition

substitute(X, P, Y, Q)

is resolved the variable Y is uninstantiated and consequently occurs uninstantiated also in the "output" Q.

Notice that in the correct metalevel representation, object-level variables are presented by variable-free terms - in this case by the terms var(1) and var(2). Naming object-level variables by variable-free terms at the metalevel is standard logical practice, and is featured in the amalgamated logic of Bowen and Kowalski [5] and in the metalogical programming language Gödel [8].

3.2   Meta for knowledge assimilation

Although there are situations where it is possible to omit the theory argument from the demo predicate altogether, there are other situations where the theory argument plays an essential role. This is illustrated in the following (simplified) representation of knowledge assimilation. Here the predicate

assimilate(T, Input, NewT)

expresses that a new theory named NewT results from assimilating an input sentence named Input into a current theory named T. The four clauses correspond approximately to the four cases of knowledge assimilation presented in section (1).

$$assimilate(T, Input, T) \leftarrow demo(T, Input)$$
$$assimilate(T, Input, NewT) \leftarrow T \equiv S \wedge T',$$
$$demo(T' \wedge Input, S),$$
$$assimilate(T', Input, NewT)$$
$$assimilate(T, Input, T) \leftarrow demo(T \wedge Input, \Box)$$
$$assimilate(T, Input, NewT) \leftarrow NewT \equiv T \wedge \Delta,$$
$$demo(NewT, Input)$$

Here in contrast with [5] and [6] I have used the conjunction symbol to combine a theory (regarded as a conjunction of sentences) with a sentence. The infix function symbol $\equiv$ can be regarded as expressing logical equivalence. This gives an elegant (but potentially inefficient) way of expressing in the second clause that

S∈T

where T is regarded as a set of clauses. The symbol □ in the third clause denotes logical contradiction. For simplicity's sake this clause caters only for the case where the input is rejected when it is inconsistent with the current theory. The fourth clause formalises a simplified form of abduction without integrity checking.

3.3   Meta for theory construction

Knowledge assimilation can be thought of as a special case of constructing new theories from old ones. Such more general theory construction is an important part of software engineering, promoting modularity, reuse of software, and programming-in-the-large. Theory construction, by its very nature, is essentially a metalinguistic operation. Applications of metalogic to theory construction are presented by Brogi et al [6] in their contribution to this Symposium.

As pointed out in [6] programming by rules and exceptions can be regarded as combining one theory "Rules" of rules together with another theory "Except" of exceptions. In the special case where the theory of exceptions consists only of unit clauses, or is logically equivalent to unit clauses, the combination has an especially simple definition. Here the function symbol "combine" names the resulting new theory

demo(combine(Rules, Except), ¬ P) ← demo(Except, ¬ P)
demo(combine(Rules, Except), P) ← demo(Rules, P),
                                   not demo(Except, ¬ P)

In the general case, where the theory of exceptions is defined in terms of predicates defined by the theory of rules, the metatheoretic definition is more complex, and begins to approximate the model theoretic definition.

3.4   Reflection

In those cases where the theory and proposed conclusion are fully given, it is possible to execute calls to the demo predicate without using a metainterpreter, by using the object level theorem-prover directly instead. This corresponds to the use of a "reflection" rule of inference

$$T \vdash P$$
$$\overline{}$$
$$demo(T', P')$$

where T' and P' name T and P respectively. T' and P' must not contain (meta) variables, because otherwise it would not be possible to (correctly) associate unique T and P with T' and P'.

For example, given a call of the form assimilate(T, Input, NewT), where T and Input are fully instantiated, the reflection rule can be used to execute the induced calls to the demo predicate in the first three clauses of the simplified definition of assimilate. It cannot be used, however, to execute the call to demo in the fourth clause, because there the term NewT in the first argument of demo contains the uninstantiated variable Δ.

The metainterpreter, on the other hand, can be executed with any pattern of input-output. Thus a metalevel definition of an object-level theorem-prover is more powerful than the object-level theorem-prover itself. As we have just seen, for example, the object level theorem-prover might be able only to perform deduction, whereas the metainterpreter would be able to perform abduction as well.

Because the metalevel is generally more powerful than the object-level, the converse reflection rule

$$demo(T', P')$$
$$\overline{}$$
$$T \vdash P$$

is also useful, especially if the metainterpreter employs more sophisticated theorem-proving techniques than the object level proof procedure. Such reflection rules (also called "attachment" rules) were first introduced by Weyhrauch [50] in FOL. They are the sole means of metareasoning in the metalogic programming language proposed by Costantini and Lanzaroni [11].

It is important to realise that reflection rules are weaker than reflection axioms:

$$demo(T', P') \leftrightarrow T \vdash P$$

Tarski [48] showed that reflection axioms lead to inconsistency. Montague [38] and Thomason [49] showed that weaker forms of reflection axioms can also lead to inconsistency.

3.5 Theories named by constants

It is often inconvenient or inappropriate to name theories by lists of names of sentences, and more useful to name them by constant symbols. Thus, for example, we might use a constant symbol, say t1, to name the theory

$$p \leftarrow q, r$$
$$p \leftarrow s$$
$$r$$
$$s$$

by asserting the metalevel clauses

demo(t1, p ← q ∧ r)
demo(t1, q ← s)
demo(t1, r)
demo(t1, s)

The object-level query

? p

can then be represented simply by the metalevel query

? demo(t1, p)

An especially important theory is the definition of the demo predicate itself. This too can be named, like any other theory, by a list or by a constant symbol. Using a constant symbol, such as "pr", and restricting ourselves for simplicity to the propositional Horn clause case, we need to assert the metametalevel axioms

(pr5) demo(pr, demo(T, P) ← demo(T, P ← Q) ∧ demo (T, Q))
(pr6) demo(pr, demo(T, P ∧ Q) ← demo(T, P) ∧ demo(T, Q))

to express that pr contains the two axioms pr1 and pr2.

Notice that, as before, I have simplified notation and not distinguished, for example, except for context, between the predicate symbol demo and the function symbol demo, which is its name.

The clauses pr5 and pr6 do not assert explicitly that pr1 and pr2 are the only axioms of pr. Moreover, it is not clear whether or not the axioms pr5 and pr6 themselves should be axioms of pr. Assuming that they should, we would then need to assert sentences such as

(pr7) demo(pr, pr5)
(pr8) demo(pr, pr6)

where, for simplicity's sake, I have used the symbols pr5 and pr6 in place of the longer names. But then, for consistency's sake, we should also include pr7 and pr8 in pr, and so on, ad infinitum. Fortunately, if we choose to follow this route, there is a simple solution. Simply add to pr the single additional axiom

(pr9) demo(pr, demo(pr, P) ← P)

We may then take pr to consist of the axioms pr1, pr2, pr5, pr6, and pr9. We can then derive

   demo(pr, demo(pr, P)) ← demo(pr, P)

by pr1 and pr9. Using this, we can derive pr7 from pr5, pr8 from pr6, and so on.

Notice that the notation of pr9 is not entirely precise. This could be remedied by employing an axiom schema

   demo(pr, demo'(pr', p') ← p)

in place of pr9, where demo', pr', and p' name demo, pr, and p respectively, and p is any sentence name. Alternatively, we could use an axiom

   demo(pr, demo'(pr', P') ← P) ← name(P, P')

where

   name(P, P')

is a metalevel predicate relating names P to their names P'.

3.6 Meta for representing knowledge and belief

Interpreting "demo" as "believe", the axioms pr1, pr2, pr5, pr6, and pr9 of pr resemble modal formulations of belief. If naming conventions are carried out conventionally and precisely, then the metatheoretic formulation is syntactically more cumbersome than the modal formulations. Moreover, the metatheoretic formulation is potentially subject to the inconsistencies discovered by Montague [38] and Thomason [49]. Their results, however, do not apply directly to our formalisation.

Bearing in mind these cautions, the metatheoretic formulation, nevertheless, offers several advantages over the modal formulation. It avoids, in particular, one of the biggest weaknesses of the standard modal approach, namely that logically equivalent formulae can be substituted for one another in any context. This gives rise, among other problems, to the problem of omniscience: that an agent believes (or knows) all logical consequences of its explicitly held beliefs (or knowledge).

As Konolige [29] points out, interpreting belief as provability allows resource limitations to restrict the implicit beliefs that can be accessed from explicitly held beliefs. Resource limitations can be captured in metatheoretical formulations by adding an extra parameter to the demo predicate to indicate the number of steps in a proof. For example

(pr1') demo(T, P, N + M) ← demo(T, P ← Q, N),
                    demo(T, Q, M)

(pr2') demo(T, P ∧ Q, N + M) ← demo(T, P, N),
                    demo(T, Q, M)

The metatheoretic approach is also more expressive than the modal approach. Even such simple sentences as pr1 and pr2, which quantify (implicitly) over names T of theories and names P and Q of sentences, are not expressible in standard modal logics. In modal logic the analogues of pr1 and pr2 are axiom schemas. Des Revières and Levesque [13] show that it is exactly this greater expressive power of metalogic that accounts for the inconsistencies that can be obtained when modal logics are reformulated in metalogical terms. They show that the inconsistencies can be avoided if the metalogical formulations are restricted analogously to the modal formulations.

Most systems of metalogic, such as FOL [50] and Gödel [8], separate object language and metalanguage. For some applications, such as the formalisation of legislation or the representation of knowledge and belief, it is convenient to combine them within a single language. Such an amalgamation of object language and metalanguage for logic programming was presented by Bowen and Kowalski [5].

The two-argument demo predicate, with amalgamation and theories named by constants, seems to be especially convenient for representing multi-agent knowledge and belief. Suppose, for example that we use the constant symbol "john", depending on the context, to name both the individual John and the theory consisting of John's beliefs. That John believes that Mary loves him could then be expressed by the metasentence

demo(john, loves(mary, john))

That John *knows* Mary loves him could be expressed by the combination of object level and metalevel sentences

(j1)   demo(john, loves(mary, john))
(j2)   loves(mary, john)

Here, as earlier, where the context makes the intended distinctions clear, I use the same notation for sentences and their names. Notice that the two sentences j1 and j2 could belong to the theory named john, the theory named mary, or any other theory.

Multi-agent knowledge and belief can also be represented in a metalogic where object languages and metalanguages are separated, as in the formalisation of the wise man problem by Aiello et al [1]. In this formulation of the problem, reflection rules are used to link different object level theories with the metatheory. The formulation of the same problem by Kim and Kowalski [25] using amalgamation of object language and metalanguage is closer to modal solutions [27, 28].

Another application where amalgamation seems to be especially appropriate is the formalisation of legislation, where one act (represented by a theory) often refers to other acts. The latest British Nationality Act 1981 [21, 46], for example, refers frequently to consequences of the previous Nationality Acts and of the Immigration Act 1971. It also contains ingenious examples of theory construction, such as that on page 13 [21], where a person is defined to be a British citizen "by descent", if that person

"... was a person who, under any provision of the British Nationality Acts 1948 to 1965, was deemed for the purposes of the proviso to section 5(1) of the 1948 Act to be a citizen of the United Kingdom and Colonies by descent only, or would have been so deemed if male".

3.7   Semantic Issues

Because of inconsistencies, such as Tarski's formalisation of the liar paradox, which can arise when object language and metalanguage are combined in the same language, most formal systems, such as FOL [50] and the proposed language Gödel [8] keep the languages separate.   However, applications of amalgamation and of the analogous modal logics demonstrate the utility of combining the different levels of language in the same formalism.   The recent work of des Revières and Levesque [13] and of Perlis [41] suggests that the semantic foundations of modal logic can also be used to underpin the amalgamation of object level and metalevel logic.   Given the link that has been established between the semantics of negation by failure and the stable expansions of autoepistemic logic [17], it may be that some extension of the semantics of autoepistemic logic [39] might provide such an underpinning.

However, it may be, instead, that Gödel's method of formalising the proof predicate of arithmetic within arithmetic will provide a better semantics for amalgamation, as was the original intention in [5].   Moreover, as several authors [4, 29, 47] have shown, metalogic itself can provide a semantics for certain modal logics.   In this approach the modal logics might be viewed as abstractions which avoid the naming complications of metalogic.

4.   Is full first-order logic necessary?

It is well known that Horn clause logic is a basis for all computation.   This means that any logic at all, with recursively enumerable axioms and effective rules of inference, can be implemented using a Horn clause metainterpreter, by means of rules such as

$$demo(T, P) \leftarrow demo(T, P \vee Q),$$
$$demo(T, \neg Q)$$

for example.   This means that knowledge that can be represented in any logic can be represented, indirectly at least, in Horn clause logic alone.

But such purely theoretical results alone are not satisfactory in practice, where we are interested in natural and direct representations of knowledge.   It is for this reason in fact that in the earlier sections of this paper I have argued that Horn clause logic programming should be augmented with such extensions as abduction, integrity constraints, and metalevel reasoning.   I now want to argue that the extension to full first-order logic may be unnecessary.

But first it is important to clarify that extended logic programming form, as identified by Lloyd and Topor [34] for example, includes the use of full first-order logic for queries, integrity constraints, and the conditions of clauses. This assymetric use of full first-order logic is exemplified most dramatically in the case of relational databases, where it has been found useful in practice to limit severely the form of assumptions in the database to variable-free atomic formulae, but to allow queries and integrity constraints in full first-order form.

Lloyd and Topor have shown that logic programs extended to allow full first-order queries, integrity constraints, and conditions of clauses can be transformed into normal logic programming form. The correctness of their transformation is based on interpreting the semantics of negation by failure as classical negation in the completion of the program. It is not entirely clear how correctness is affected if some other semantics is taken instead.

The starting point for my argument that unrestricted, full first-order logic may not be needed for knowledge representation is an empirical claim: Most published examples of the use of logic in computing and artificial intelligence use logic programming form or some modest extension of logic programming form, usually without explicitly recognising it. I have already referred to several such examples in section (2). The Yale shooting problem, for example, is formulated directly in logic programming form. The birds-fly example is expressed in one formulation directly as a logic program and in another formulation as a logic program with exceptions. Similarly, the quaker-republican example has natural formulations both as a logic program with abduction and integrity constraints and as a logic program with exceptions. Numerous other examples can be found in the literature.

Another major class of applications is the formalisation of legislation [33]. The form of natural language in which legislation is written has a marked similarity to logic programming form. It also exemplifies many of the extensions, such as exceptions having negative conclusions [31] and metalinguistic constructs, which are needed for other purposes.

Interestingly, the natural language formulation of legislation generally expresses only the if-halves of definitions, and only rarely uses the if-and-only-if form. I have not found a single example of a sentence with a disjunctive conclusion.

Examples of disjunctive conclusions can, however, be found elsewhere in computing and artificial intelligence. It seems that many of these examples can often usefully be reformulated in an extended logic programming form.

I have already mentioned in section (2) that the deductive, non-Horn formulation of abduction expresses the only-if half of the if-and-only-if completion of the explicitly abductive, Horn formulation. As Clark showed in the case of negation by failure [9], reasoning about all derivations determined by a Horn clause program often simulates reasoning with the only-if, non-Horn, half of the completion of the program.

Some disjunctions

P v Q

can often be reexpressed naturally as one or two clauses in logic programming form

P ← not Q
Q ← not P

In the event calculus [30], for example, we found it useful to express the disjunction

for all time intervals I1 and I2 and properties P,
if P holds maximally for I1
and P holds maximally for I2,
then I1 = I2 or I1 preceeds I2 or I2 preceeds I1

in the logic programming form

I1 = I2 ← P holds maximally for I1,
P holds maximally for I2,
not I1 preceeds I2,
not I2 preceeds I1.

We also found it useful to reexpress the existential quantifier in the sentence

for every event type E and property P,
there exists a time interval I, such that,
if E initiates P,
then E starts I
and P holds maximally for I

by means of a function symbol which names I as a function of E and P

> E starts after(E, P) ← E initiates P
> P holds maximally for after(E, P) ← E initiates P.

The extent to which it is natural to use logic programming for knowledge representation might also be related to arguments for using intuitionistic logic for mathematics. There are several parallels between extensions of logic programming and intuitionistic logic. To begin with, the Horn clause basis of logic programming is a sublanguage of both classical and intuitionistic logic, and negation by failure is a constructive form of negation, in the spirit of intuitionistic negation. Moreover, the lack of disjunctive conclusions and existential statements in logic programming is similar to the property of intuitionistic logic that

> if a disjunction P v Q is provable
> then P is provable or Q is provable, and

> if an existential statement ∃XP(X) is provable then
> P(t) is provable for some term t.

There are cases, however, where reasoning with disjunctive conclusions seems unavoidable, even when starting from a logic programming position. Such a case arises, for example, when trying to prove that a certain property of a deductive database holds, independently of specific data.

The following very simplified example comes from the Alvey Department of Health and Social Security demonstrator project [3]. The clauses of the theory represent possible legislation and associated socio-economic hypotheses.

> receives(X, £20) ← adult(X), age(X) $\geq$ 65
> receives(X, £25) ← adult(X), age(X) < 65, unemployed(X)
> receives(X, wages(X)) ← age(X) < 65, adult(X), employed(X)
> £25 $\leq$ wages(X) ← adult(X)
> adequate-income(X) ← receives(X, Y), £20 $\leq$ Y
> X $\leq$ Y ← X $\leq$ Z, Z $\leq$ Y

Given such a theory, it might be desirable to show that, as a consequence, every adult has an adequate income

> adequate-income(X) ← adult(X)

Unfortunately, even though the theory and theorem can be expressed in logic programming form, the theorem cannot be proved without extra disjunctive assumptions, such as

age(X) < 65 v age(X) ≥ 65 ← adult(X)
employed(X) v unemployed(X) ← adult(X)

But contrary to first appearances, the problem does not necessary go beyond logic programming form. The additional assumptions can be viewed as integrity constraints that constrain the information that can be added to the theory, rather than as statements belonging to the theory. Nonetheless, the use of a non-Horn clause theorem-prover, such as Satchmo [37], to solve the problem seems to be unavoidable. How this use of an object-level formulation of integrity constraints can be reconciled with arguments for a metalevel or modal formulation has still to be resolved.

Conclusion

I have argued in this paper that certain extensions of logic programming are needed for knowledge representation and knowledge assimilation. These extensions include abduction, integrity constraints, and metalevel reasoning. On the other hand, I have also argued that other extensions may not be needed - extensions such as modal logic and full first-order logic.

I have touched upon some of the unresolved semantic issues concerning these extensions, arguing for a conservative approach, which builds upon and does not complicate the semantics of the underlying logic. But these are complicated technical matters, about which it is good to have an open mind.

References

[1]   Aiello, L. C., Nardi, D. and Schaerf, M. [1988]: "Reasoning about Knowledge and Ignorance", Proceedings of the FGCS, pp. 618-627.

[2]   Apt, K. R., and Bezem, M. [1990]: "Acyclic programs", Proc. of the Seventh International Conference on Logic Programming, MIT Press, pp. 579-597.

[3]   Bench-Capon, T.J.M. [1987]: "Support for policy makers: formulating legislation with the aid of logical models", Proc. of the First International Conference on AI and Law, ACM Press, pp. 181-189.

[4]   Boolos, G. [1979]: The Unprovability of Consistency, Cambridge University Press.

[5]   Bowen, K. A. and Kowalski, R. A. [1982]: "Amalgamating Language and Metalanguage in Logic Programming", in Logic Programming (Clark, K.L. and Tärnlund, S.-Å., editors), Academic Press, pp. 153-173.

[6]   Brogi, A., Mancarella, P., Pedreschi, D., Turini, F. [1990]: "Composition operators for logic theories", Proc. Symposium on Computational Logic, Springer-Verlag.

[7]   Bry, F., "Intensional updates: abduction via deduction". Proceedings of the Seventh International Conference on Logic Programming, MIT Press, pp. 561-575.

[8]   Burt, A. D., Hill, P. M. and Lloyd, J. W. [1990]: "Preliminary Report on the Logic Programming Language Gödel. University of Bristol, TR-90-02.

[9]   Clark, K. L. [1978]: "negation by failure", in "Logic and databases", Gallaire, H. and Minker, J. [eds], Plenum Press, pp. 293-322.

[10]   Console, L., Theseider Dupré, D., and Torasso, P. [1990]: "A completion semantics for object-level deduction", Proc. AAAI Symposium on Automated Abduction, Stanford, March 1990.

[11]   Costantini, S. and Lanzarone, G. A. [1989]: "A metalogic programming language", Proc. Sixth International Conference on Logic Programming, pp.218-233.

[12]   Cox, P. T. and Pietrzykowski, T. [1986]: "Causes for Events: Their Computation and Applications", in Proceedings CADE-86, pp 608-621.

[13]   des Rivières, J. and Levesque, H. J. [1986]: "The Consistency of Syntactical Treatments of Knowledge", in Proceedings of the 1986 Conference on Theoretical Aspects of Reasoning about Knowledge, (Halpern, J. editor), pp. 115-130.

[14]   Eshghi, K. and Kowalski, R. A. [1989]: "Abduction compared with negation by failure", Proceedings of the Sixth International Logic Programming Conference, MIT Press, pp. 234-255.

[15]   Evans, C. [1989]:   "Negation-as-failure as an approach to the Hanks and McDermott problem", Proc. Second International Symposium on Artificial Intelligence, Monterrey, Mexico, 23-27 October 1989.

[16]   Finger, J. J. and Genesereth, M.R. [1985]: "RESIDUE: A Deductive Approach to Design Synthesis", Stanford University Report No. CS-85-1035.

[17]   Gelfond, M. and Lifschitz, V. [1988]: "The stable model semantics for logic programs", Proceedings of the Fifth International Conference and Symposium on Logic Programming, (Kowalski, R. A. and Bowen, K. A. editors), volume 2, pp. 1070-1080.

[18]   Gelfond, M. and Lifschitz, V. [1990]: "Logic programs with classical negation", Proceedings of the Seventh International Conference on Logic Programming, MIT Press, pp. 579-597.

[19]   Hanks, S. and McDermott, D. [1986]: "Default reasoning, non-monotonic logics, and the frame problem", Proc. AAAI, Morgan and Kaufman, pp. 328-333.

[20]   Hill, P. M. and Lloyd, J. W. [1989]: "Analysis of metaprograms", In Metaprogramming in Logic Programming", (H.D. Abramson and M.H. Rogers, editors), MIT Press, pp. 23-52.

[21]   H.M.S.O. [1981]: "British Nationality Act 1981", Her Majesty's Stationery Office, London.

[22]   Kakas, A. C. and Mancarella, P. [1989], "Anomalous models and Abduction", in Proceedings of 2nd International Symposium on Artificial Intelligence, Monterrey, Mexico, 23-27 October 1989.

[23]   Kakas, A. C. and Mancarella, P. [1990], "Generalised stable models: a semantics for abduction", Proceedings of ECAI 90, pp. 385-391.

[24]   Kakas, A. C. and Mancarella, P. [1990], "Database updates through abduction". Proceedings of VLDB 90.

[25]   Kim, J.-S., and Kowalski, R. A. [1990], "An application of amalgamated logic to multi-agent belief", Proceedings of Meta 90, Leuven University.

[26]   Konolige, K. [1982]: "A first order formalization of knowledge and action for a multiagent planning system", in Machine Intelligence 10, (Hayes, J.E., Michie, D., Pao, Y. H., editors), Ellis Horwood.

[27]   Konolige, K. [1982]: "Circumscriptive Ignorance", Proc. AAAI-82, pp. 202-204.

[28]   Konolige, K. [1985]: "Belief and Incompleteness", in Formal Theories of the Commonsense World, (Hobbs, J. and Moore, R. C. editors), Ablex Pub. Corp., pp. 359-403.

[29]   Konolige, K. [1986]: "A Deduction Model of Belief", Pitman Research Notes in Artificial Intelligence.

[30]   Kowalski, R. A. and Sergot, M. J. [1986]: "A logic-based calculus of events", New Generation Computing, Vol. 4, No. 1, pp. 67-95.

[31]   Kowalski, R. A. [1989]: "The treatment of negation in logic programs for representing legislation", Proceedings of the Second International Conference on Artificial Intelligence and Law, pp. 11-15.

[32]   Kowalski, R. A. and Sadri, F. [1990], "Logic programs with exceptions", Proceedings of the Seventh International Conference on Logic Programming, MIT Press, pp. 598-613.

[33]   Kowalski, R. A., Sergot, M. J. [1990]: "The use of logical models in legal problem solving", Ratio Juris, Vol. 3, No. 2, pp. 201-218.

[34]   Lloyd, J. W. and Topor, R. W. [1984]: "Making Prolog more expressive", Journal of Logic Programming, Vol. 3, No. 1, pp. 225-240.

[35]   Lloyd J. W. [1987]: "Foundations of logic programming", second extended edition, Springer-Verlag.

[36]   McCarthy, J. [1980]: "Circumscription - a form of nonmonotonic reasoning", Artificial Intelligence, Vol. 26, No. 3, pp. 89-116.

[37]   Manthey, R. and Bry, F. [1988]: "SATCHMO: A theorem prover implemented in Prolog", Proc. Nineth Int. Conf. on Automated Deduction (CADE).

[38]   Montague, R. [1963]: "Syntactical Treatments of Modality, with Corollaries on Reflection Principles and Finite Axiomatizability", Acta Philosophic Fennica 16, pp. 153-167.

[39]   Moore, R.C. [1985]: "Semantical Considerations on Nonmonotonic Logic in Artificial Intelligence, Artificial Intelligence 25, (1985), pp. 75-94.

[40]   Pereira, L. M. and Aparicio, J. N. [1990]: "Default reasoning as abduction", Technical report, AI Centre/Uninova, 2825 Monte da Caparica, Portugal.

[41]   Perlis, D. [1988]: "Language with Self-Reference II: Knowledge, Belief and Modality", Artificial Intelligence 34, pp. 179-212.

[42]   Poole, D. [1988]: "A logical framework for default reasoning", Artificial Intelligence 36, pp. 27-47.

[43]   Reiter, R. [1980]: "A logic for default reasoning", Artificial Intelligence, 13 (1,2), pp. 81-132.

[44]   Reiter, R. [1990]: "On asking what a database knows", Proc. Symposium on Computational Logic, Springer-Verlag.

[45]   Sadri, F. and Kowalski, R. A. [1987]: "A theorem proving approach to database integrity", In Foundations of deductive databases and logic programming (J. Minker, editor), Morgan Kaufmann, pp. 313-362.

[46]   Sergot, M. J., Sadri, F., Kowalski, R. A., Kriwaczek, F., Hammond, P. and Cory, H. T. [1986]: " The British Nationality Act as a logic program", CACM, Vol. 29, No. 5, pp. 370-386.

[47]   Smorynski, C. [1985], "Self-Reference and Modal Logic", Springer-Verlag, New York.

[48]   Tarski, A. [1936]: "Der Wahrheitsbegriff in den formalisierten Sprachen", Studia Philosophia, Vol. 1, pp. 261-405. English translation "The concept of truth in formalised languages" in A. Tarski "Logic, Semantics, and Metamathematics, Clarendon, Oxford, 1956.

[49]   Thomason, Richmond H., [1980]: "A note on syntactic treatments of modality",
        Synthese, Vol. 44, pp. 391-395.

[50]   Weyhrauch, R. W. [1980]: "Prolegomena to a Theory of Mechanized Formal
        Reasoning", Artificial Intelligence 13, pp. 133-170.

[51]   Charniak, E. and McDermott, D. [1985]: "Introduction to Artificial Intelligence",
        Addison-Wesley.

AN INTRODUCTION TO PROLOG III

Alain Colmerauer
Groupe Intelligence Artificielle, Faculté des Sciences de Luminy,
Case 901, 70 route Léon Lachamp, 13288 Marseille Cedex 9, France.

Abstract. The Prolog III programming language extends Prolog by redefining the fundamental process at its heart : unification. Into this mechanism, Prolog III integrates refined processing of trees and lists, number processing, and processing of two-valued Boolean algebra. We present the specification of this new language and illustrate its capabilities by means of varied examples. We also present the theoretical foundations of Prolog III, which in fact apply to a whole family of programming languages. The central innovation is to replace the concept of unification by the concept of constraint solving.

INTRODUCTION

Prolog was initially designed to process natural languages. Its application in various problem solving areas has brought out its qualities, but has also made clear its limits. Some of these limitations have been overcome as a result of more and more efficient implementations and ever richer environments. The fact remains, however, that the core of Prolog, namely, Alan Robinson's unification algorithm [22], has not fundamentally changed since the time of the first Prolog implementations, and is becoming less and less significant compared to the ever-increasing number of external procedures as, for example, the procedures used for numerical processing. These external procedures are not easy to use. Their invocation requires that certain parameters are completely known, and this is not in line with the general Prolog philosophy that it should be possible anywhere and at any time to talk about an unknown object x.

In order to improve this state of affairs, we have fundamentally reshaped Prolog by integrating at the unification level : (1) a refined manipulation of trees, including infinite trees, together with a specific treatment of lists, (2) a complete treatment of two-valued Boolean algebra, (3) a treatment of the operations of addition, subtraction, multiplication by a constant and of the relations $<, \leq, >, \geq$, (4) the general processing of the relation $\neq$. By doing so we replace the very concept of unification by the concept of constraint solving in a chosen mathematical structure. By mathematical structure we mean here a domain equipped with operations and relations, the operations being not necessarily defined everywhere.

The result of incorporating the above features into Prolog is the new programming language Prolog III. In this paper[1] we establish its foundations and illustrate its capabilities using representative examples. These foundations, which in fact apply to a whole family of "Prolog III like" programming languages, will be presented by means of simple mathematical concepts without explicit recourse to first-order logic.

The research work on Prolog III is not an isolated effort; other research has resulted in languages whose design shares features with Prolog III. The CLP(R) language developed by J. Jaffar and S. Michaylov [19] emphasizes real number processing, whereas the CHIP language developed by the team led by M. Dincbas [13] emphasizes processing of Boolean algebra and pragmatic processing of integers and elements of finite sets. Let us also mention the work by J. Jaffar et J-L. Lassez [18] on a general theory of "Constraint Logic Programming". Finally, we should mention Prolog II, the by now well-established language which integrates infinite trees and the ≠ relation, and whose foundations [8,9] were already presented in terms of constraint solving. From a historical point of view, Prolog II can thus be regarded as the first step towards the development of the type of languages discussed here.

THE STRUCTURE UNDERLYING PROLOG III

We now present the particular structure which is the basis of Prolog III and specify the general concept of a structure at the same time. By *structure* we mean a triple (D, F, R) consisting of a *domain* D, a set F of *operations* and a set of relations on D.

Domain

The domain D of a structure is any set. The domain of the structure chosen for Prolog III is the set of *trees* whose nodes are labeled by one of the following :

(1) identifiers,
(2) characters,
(3) Boolean values, 0' and 1',
(4) real numbers,
(5) special signs $<>^\alpha$, where α is either zero or a positive irrational number.

Here is such a tree :

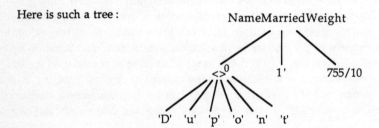

---

[1] This paper has been published with the same title in the Communications of the ACM 33, 7 (July 1990). The Association for Computing Machinery, who owns the copyright, gave permission for copying. A very preliminary version of the paper has appeared in the *Proceedings of the 4th Annual ESPRIT Conference*, Brussels, North Holland, pp. 611-629, 1987.

The branches emanating from each node are ordered from left to right; their number is finite and independent of the label attached to the node. The set of nodes of the tree can be infinite. We do not differentiate between a tree having only one node and its label. Identifiers, characters, Boolean values, real numbers and special signs $<>^{\alpha}$ will therefore be considered to be particular cases of trees.

By *real numbers* we mean perfect real numbers and not floating point numbers. We make use of the partition of the reals into two large categories, the rational numbers, which can be represented by fractions (and of which the integers are a special case) and the irrational numbers (as for example π and $\sqrt{2}$) which no fraction can represent. In fact, the machine will compute with rational numbers only and this is related to an essential property of the constraints that can be employed in Prolog III : if a variable is sufficiently constrained to represent a unique real number then this number is necessarily a rational number.

A tree a whose initial node is labeled by $<>^{\alpha}$ is called a *list* and is written
$$<a_1,...,a_n>^{\alpha},$$

where $a_1...a_n$ is the (possibly empty) sequence of trees constituting the immediate daughters of a. We may omit α whenever α is zero. The *true* lists are those for which α is zero : they are used to represent sequences of trees (the sequence of their immediate daughters). Lists in which α is not zero are *improper* lists that we have not been able to exclude : they represent sequences of trees (the sequence of their immediate daughters) completed at their right by something unknown of length α. The *length* $|a|$ of the list a is thus the real $n+\alpha$. True lists have as their length a non-negative integer and improper lists have as their length a positive irrational number. The list $<>$ is the only list with length zero, it is called the *empty list*. We define the operation of *concatenation* on a true list and an arbitrary list by the following equality :

$$<a_1,...,a_m>^0 \bullet <b_1,...,b_n>^{\alpha} = <a_1,...,a_m,b_1,...,b_n>^{\alpha}.$$

This operation is associative, $(a \bullet a') \bullet b = a \bullet (a' \bullet b)$, and the empty list play the role of the neutral element, $a \bullet <> = a$ et $<> \bullet b = b$. We observe that for any list b there exists one and only one true list a and one and only one real α such that

$$b = a \bullet <>^{\alpha}.$$

This list a is called the *prefix* of b and is written $\lfloor b \rfloor$.

Let D^n denote the set of tuples $a_1...a_n$ constructed on the domain D of a structure. An *n-place operation f* is a mapping from a subset E of D^n to D,

$$f: a_1...a_n \mapsto f a_1...a_n .$$

Note that if E is strictly included in D^n, the operation f is partial; it is not defined for all tuples of size n. Note also that in order to be systematic the result of the operation is written in prefix notation. The 0-place operations are simply mappings of the form

$$f: \wedge \mapsto f,$$

where $\wedge$ is the empty tuple; they are also called *constants* since they can be identified with elements of the domain.

As far as the chosen structure is concerned here is first of all a table listing the operations which belong to F. In this table we introduce a more graceful notation than the prefix notation.

Constants

id	:	$\wedge \mapsto$ id,
'c'	:	$\wedge \mapsto$ 'c',
0'	:	$\wedge \mapsto$ 0',
1'	:	$\wedge \mapsto$ 1',
q	:	$\wedge \mapsto q$,
$<>^0$	:	$\wedge \mapsto <>$,
$c_1...c_m$	:	$\wedge \mapsto$ "$c_1...c_m$".

Boolean operations

$\neg$	:	$b_1 \mapsto \neg b_1$,
$\wedge$	:	$b_1 b_2 \mapsto b_1 \wedge b_2$,
$\vee$	:	$b_1 b_2 \mapsto b_1 \vee b_2$,
$\supset$	:	$b_1 b_2 \mapsto b_1 \supset b_2$,
$\equiv$	:	$b_1 b_2 \mapsto b_1 \equiv b_2$.

Numerical operations

$+^1$	:	$r_1 \mapsto +r_1$,
$-^1$	:	$r_1 \mapsto -r_1$,
$+^2$	:	$r_1 r_2 \mapsto r_1 + r_2$,
$-^2$	:	$r_1 r_2 \mapsto r_1 - r_2$,
$q\times$	:	$r_1 \mapsto q \times r_1$,
$/q'$	:	$r_1 \mapsto r_1 / q'$.

List operations

$\|$	:	$l_1 \mapsto |l_1|$,
$<,>^m$	:	$a_1...a_m \mapsto <a_1,...,a_m>$,
$a_1...a_n \bullet$	:	$l_1 \mapsto <a_1,...,a_n> \bullet l_1$.

General operations

$()^{n+2}$	:	$e_1 a_2...a_{n+2} \mapsto e_1(a_2,...,a_{n+2})$,
$[]$	:	$e_1 l_2 \mapsto e_1[l_2]$.

Here *id* designates an identifier, c and c_i a character, q et q' rational numbers represented by fractions (or integers), with q' not zero, m a positive integer, n a non-negative integer and a_i an arbitrary tree. The result of the different operations is defined only if b_i is a Boolean value, r_i a real number, l_i a list and e_i a label not of the form $<>^\alpha$.

To each label corresponds a constant, with the exception of irrational numbers and labels of the form $<>^\alpha$, where α is not zero. The constant "$c_1...c_m$" designates the true list whose immediate daughters make up the sequence of characters 'c_1' ... 'c_m'. The operations $\neg, \wedge, \vee, \supset, \equiv$, correspond to the classical Boolean operations when they are defined. The operations $\pm^1, \pm^2, q\times$, when they are defined, are the 1-place $\pm$, the 2-place $\pm$, multiplication by the constant q (when this does not lead to confusion we may omit the sign ×) and division by the constant q'. By $|l_1|$ we designate the length of the list l_1. By $<a_1,...,a_m>$ we designate the true list whose immediate daughters make up the sequence $a_1,...,a_m$. The operation $a_1...a_n \bullet$ applied to a list l_1 consists in concatenating the true list $<a_1,...,a_n>$ to the left of l_1. By $e_1(a_2,...,a_{n+2})$ we designate the tree consisting of an initial node labeled e_1 and the sequence of immediate daughters $a_2,...,a_{n+2}$. By $e_1[l_2]$ we designate the tree consisting of an initial node labeled e_1 and of the sequence of immediate daughters of the list l_2.

We note the following equalities (provided the different operations used are indeed defined) :

$$"c_1...c_m" = <'c_1', ... ,'c_m'>$$

$$a_0(a_1,...,a_m) = a_0[<a_1,...,a_m>].$$

Using the constants and the operations we have introduced, we can represent our previous example of a tree by

NameMarriedWeight("Dupont", 1', 755/10)

or by

NameMarriedWeight[<<'D','u'>•"pont", 0'∨1', 75+1/2>].

Relations

Let D^n again denote the set of tuples $a_1...a_n$ constructed on the domain D of a structure. An *n-place relation* r is a subset E of D^n to D. To express that the tuple $a_1...a_n$ is in the relation r we write

$$r\, a_1...a_n .$$

With respect to the structure chosen for Prolog III, here are the relations contained in F. We also introduce a more graceful notation than the prefix notation.

One-place relations

id   : a_1 : id,
char : a_1 : char,
bool : a_1 : bool,
num : a_1 : num,
irint : a_1 : irint,
list : a_1 : list,
leaf : a_1 : leaf.

Identity relations

= : $a_1 = a_2$,
≠ : $a_1 \neq a_2$.

Boolean relations

⇒   : $a_1 \Rightarrow a_2$.

Numerical relations

<   : $a_1 < a_2$,
>   : $a_1 > a_2$,
≤   : $a_1 \leq a_2$,
≥   : $a_1 \geq a_2$,

Approximated operations

$/^3$: $a_3 \doteq a_1 \; / \; a_2$,
$\times^{n+1}$: $a_{n+1} \doteq a_1 \times ... \times a_n$,
$\bullet^{n+1}$: $a_{n+1} \doteq a_1 \bullet ... \bullet a_n$.

Here n designates an integer greater than 1 and a_i an arbitrary tree. The relations id, char, bool, num, irint, list and leaf are used to specify that the tree a_1 is an identifier, a character, a Boolean value, a real number, an integer or irrational number, a list, a label not of the form $<>^\alpha$. The relations = and ≠ correspond of course to the equality and inequality of trees. The pair of trees $a_1 a_2$ is in the relation ⇒ only if a_1 et a_2 are Boolean values and if $a_1 = 1'$ entails that $a_2 = 1'$. The pair of trees $a_1 a_2$ is in relation <, >, ≤, ≥ only if it is a pair of reals in the corresponding classical relation.

We use the relation $/^3$ to approximate division and write

$$a_3 \doteq a_1 \; / \; a_2$$

to express, on the one hand, that a_1, a_2 and a_3 are real numbers, with a_2 not equal to zero and, on the other hand, that if at least one of the reals a_2 et a_3 is rational, it is true that

$$a_3 = a_1/a_2.$$

We use the relations $\times^{n+1}$, with $n \geq 2$, to approximate a series of multiplications and write

$$a_{n+1} \doteq a_1 \overset{\cdot}{\times} ... \overset{\cdot}{\times} a_n$$

to express, on the one hand, that the a_i's are real numbers and, on the other hand, that if the sequence $a_1...a_n$ contains n or $n-1$ rationals numbers, it is true that

$$a_{n+1} = a_1 \times ... \times a_n.$$

We use the relations $\bullet^{n+1}$, with $n \geq 2$, to approximate a series of concatenations and write

$$a_{n+1} \doteq a_1 \overset{\cdot}{\bullet} ... \overset{\cdot}{\bullet} a_n$$

to express that in all cases the a_i's are lists such that

$$|a_{n+1}| = |a_1| + ... + |a_n|$$

and that, according as the element $a_1 \bullet ... \bullet a_n$ is, or is not defined,

$$a_{n+1} = a_1 \bullet ... \bullet a_n$$

or

a_{n+1} is of the form $\lfloor a_1 \bullet ... \bullet a_k \rfloor \bullet b,$

where b is an arbitrary list and k is the largest integer such that the element $a_1 \bullet ... \bullet a_k$ is defined.

We recall that $a_1 \bullet ... \bullet a_k$ is defined only if the lists $a_1, ..., a_{k-1}$ are all true lists. We also recall also that $\lfloor a \rfloor$ designates the prefix of a, that is to say, the true list obtained by replacing the initial label $<>^\alpha$ of a with the label $<>^0$.

Terms and constraints

Let us suppose that we are working in a structure (D, F, R) and let V be a *universal* set of variables, given once and for all, used to refer to the elements of its domain D . We will assume that V is infinite and countable. We can now construct syntactic objects of two kinds, terms and constraints. *Terms* are sequences of juxtaposed elements from $V \cup F$ of one of the two forms,

$$x \quad \text{or} \quad f t_1 ... t_n,$$

where x is a variable, f an n-place operation and where the t_i's are less complex terms. *Constraints* are sequences of juxtaposed elements from $V \cup F \cup R$ of the form

$$r\ t_1...t_n,$$

where r is an n-place relation and the t_i's are terms. We observe that in the definition of terms we have not imposed any restriction on the semantic compatibility between f and the t_i's. These restrictions, as we will see, are part of the mechanism which takes a term to its "value".

We introduce first the notion of an *assignment* σ to a subset W of variables : such an assignment is simply a mapping from W into the domain D of the structure. This mapping σ extends naturally to a mapping σ^* from a set T_σ of terms into D specified by

$$\sigma^*(x) = \sigma(x),$$
$$\sigma^*(f\ t_1...t_n) = f\,\sigma^*(t_1)...\sigma^*(t_n).$$

The terms that are not members of T_σ are those containing variables not in W, and those containing partial operations not defined for the arguments $\sigma^*(t_i)$. Depending on whether a term t belongs or does not belong to T_σ the *value* of t under the assignment σ is defined and equal to $\sigma^*(t)$ or is not defined. Intuitively, the value of a term under an assignment is obtained by replacing the variables by their values and by evaluating the term. If this evaluation cannot be carried out, the value of the term is not defined for this particular assignment.

We say that the assignment σ to a set of variables *satisfies* the constraint $r\ t_1...t_n$ if the value $\sigma^*(t_i)$ of each term t_i is defined and if the tuple $\sigma^*(t_1)...\sigma^*(t_n)$ is in the relation r, that is to say if

$$r\ \sigma^*(t_1)...\sigma^*(t_n).$$

Here are some examples of terms associated with the structure chosen for Prolog III. Instead of using the prefix notation, we adopt the notations used when the different operations were introduced.

$$<x>\bullet y,$$
$$x[y],$$
$$<x>\bullet 10,$$
$$duo(+x, x\vee y).$$

The first term represents a list consisting of an element x followed by the list y. The second term represents a tree, which is not a list, whose top node is labeled by x and whose list of immediate daughters is y. The value of the third term is never defined, since the concatenation of numbers is not possible. The value of the last term is not defined under any assignment, since x cannot be a number and a Boolean value at the same time.

Here are now some examples of constraints. Again we adopt the notations introduced together with the different Prolog III relations.

$$z = y-x,$$
$$x \wedge \neg y \Rightarrow x \vee z,$$
$$i+j+k \leq 10,$$
$$\neg x \neq y+z,$$
$$\neg x \neq y+x.$$

We observe that there exist assignments to $\{x, y, z\}$ which satisfy the next to the last constraint (for example $\sigma(x) = 0'$, $\sigma(y) = 2$, $\sigma(z) = 2$), but that there is no assignment which satisfies the last constraint (the variable x cannot be a number and a Boolean value at the same time).

Systems of constraints

Any finite set S of constraints is called a *system of constraints*. An assignment σ to the universal set V of variables which satisfies every constraint of S is a *solution* of S. If σ is a solution of S and W a subset of V, then the assignment σ' to W which is such that for every variable x in W we have $\sigma'(x) = \sigma(x)$ is called a *solution* of S *on* W. Two systems of constraints are said to be *equivalent* if they have the same set of solutions and are said to be *equivalent on* W if they have the same set of solutions on W.

We illustrate these definitions with some examples from our structure.
- The assignment σ to V where $\sigma(x) = 1'$ for every variable x is a solution of the system of constraints $\{x = y, y \neq 0\}$, but it is not a solution of the system $\{x = y, +y \neq 0\}$.
- The assignment σ to $\{y\}$ defined by $\sigma(y) = 4$ is a solution on $\{y\}$ of the system $\{x = y, y \neq 0\}$.
- The systems $\{x = y, +y \neq 0\}$ and $\{-x = -y, y \neq 0\}$ are equivalent. Similarly, the system $\{1 = 1, x = x\}$ is equivalent to the empty constraint system.
- The systems $\{x \leq y, y \leq z, x \neq z\}$ and $\{x < z\}$ are not equivalent, but they are equivalent on the subset of variables $\{x, z\}$.

It should be noted that all solvable systems of constraints are equivalent on the empty set of variables, and that all the non-solvable systems are equivalent. By *solvable* system, we of course mean a system that has at least one solution.

The first thing Prolog III provides is a way to determine whether a system of constraints is solvable and if so, to solve the system. For example, to determine the number x of pigeons and the number y of rabbits such that together there is a total of 12 heads and 34 legs, the following query

$$\{x+y = 12, 2x+4y = 34\} \ ?$$

gives rise to the answer

$$\{x = 7, y = 5\}.$$

To compute the sequence z of 10 elements which results in the same sequence no matter whether 1,2,3 is concatenated to its left or 2,3,1 is concatenated to its right, it will suffice to pose the query

$$\{|z| = 10, \ <1,2,3>{\bullet}z \doteq z{\bullet}<2,3,1>\} \ ?$$

The unique answer is

$$\{z = <1,2,3,1,2,3,1,2,3,1>\}.$$

If in the query the list <2,3,1> is replaced by the list <2,1,3> there is no answer, which means that the system

$$\{|z| = 10, \ <1,2,3>{\bullet}z \doteq z{\bullet}<2,1,3>\}$$

is not solvable. In these examples the lists are all of integer length and are thus true lists. As a result, approximated concatenations behave like true concatenations.

In this connection, the reader should verify that the system

$$\{<1>{\bullet}z \doteq z{\bullet}<2>\}$$

is solvable (it suffices to assign to z any improper list having no immediate daughters), whereas the system

$$\{|z| = 10, \ <1>{\bullet}z \doteq z{\bullet}<2>\},$$

which constraints z to be a true list, is not solvable. The same holds for approximated multiplication and division. Whereas the system

$$\{z \doteq x \times y, x \geq 1, y \geq 1, z < 0\}$$

is solvable (because the approximated product of two irrational numbers is any number), the system

$$\{z \doteq x \times y, x \geq 1, y \geq 1, z < 0, y \leq 1\},$$

which constrains y to be a rational number, is not solvable.

Another example of the solving of systems is the beginning of a proof that God exists, as formalized by George Boole [4]. The aim is to show that "something has always existed" using the following 5 premises :
(1) Something is.
(2) If something is, either something always was, or the things that now are have risen out of nothing.
(3) If something is, either it exists in the necessity of its own nature, or it exists by the will of another Being.
(4) If it exists by the will of its own nature, something always was.

(5) If it exists by the the will of another being, then the hypothesis, that the things which now are have risen out of nothing, is false.

We introduce 5 Boolean variables with the following meaning :
$a = 1'$ for " Something is",
$b = 1'$ for " Something always was",
$c = 1'$ for " The things which now are have risen from nothing",
$d = 1'$ for " Something exists in the necessity of its own nature ",
$e = 1'$ for "Something exists by the will of another Being".
The 5 premises are easily translated into the system

$$\{a = 1' \ a \Rightarrow b \vee c, \ a \Rightarrow d \vee e, \ d \Rightarrow b, \ e \Rightarrow \neg c\}$$

which when executed as a query produces the answer

$$\{a = 1', \ b = 1', \ d \vee e = 1', \ e \vee c = 1' \}.$$

One observes that the value b is indeed constrained to 1'.

After these examples, it is time to specify what we mean by "solving" a system S of constraints involving a set W of variables. Intuitively, this means that we have to find all the solutions of S on W. Because there may be an infinite set of such solutions, it is not possible to enumerate them all. What is however possible is to compute a system in *solved* form equivalent to S and whose "most interesting" solutions are explicitly presented. More precisely by system in *solved* form, we understand a solvable system such that, for every variable x, the solution of S on $\{x\}$ is explicitly given, whenever this solution is unique. One can verify that in the preceding examples the systems given as answers were all in solved form.

Before we end this section let us mention a useful property for solving systems of constraints in the chosen structure.

PROPERTY. If S a system of Prolog III constraints and W a set of variables, then the two following propositions are equivalent :
(1) for every x in W, there are several numerical solutions of S on $\{x\}$;
(2) there exists a numerical irrational solution of S on W.

By numerical solution, or irrational numerical solution, on a set of variables, we understand a solution in which all the variables in this set have real numbers as values, or irrational numbers as values.

SEMANTICS OF PROLOG III LIKE LANGUAGES

On the basis of the structure we have chosen, we can now define the programming language Prolog III.   As the method employed is independent of the chosen structure, we define in fact the notion of a "Prolog III like" language associated with a given structure. The only assumption that we will make is that the equality relation is included in the set of relations of the structure in question.

Meaning of a program.

In a Prolog III like language, a program is a definition of a subset of the domain of the chosen structure (the set of trees in the case of Prolog III). Members of this subset are called *admissible* elements. The set of admissible elements is in general infinite and constitutes - in a manner of speaking - an enormous hidden database. The execution of a program aims at uncovering a certain part of this database.

Strictly speaking, a program is a set of *rules*: Each rule has the form

$$t_0 \rightarrow t_1 ... t_n , S$$

where n can be zero, where the t_i's are terms and where S is a possibly empty system of constraints (in which case it is simply absent). The meaning of such a rule is roughly as follows: "provided the constraints in S are satisfied, t_0 is an admissible element if t_1 and ... and t_n are admissible elements (or if $n = 0$) ". Here is such a set of rules; this is our first example of a Prolog III program. It is an improvement on a program which is perhaps too well-known, but which remains a useful pedagogical tool : the calculation of well-balanced meals [9].

```
LightMeal(h, m, d) →
   HorsDœuvre(h, i)
   MainCourse(m, j)
   Dessert(d, k),
   {i ≥ 0, j ≥ 0, k ≥ 0, i+j+k ≤ 10};

MainCourse(m, i) → Meat(m, i);
MainCourse(m, i) → Fish(m, i);

HorsDœuvre(radishes, 1) →;
HorsDœuvre(pâté, 6) →;

Meat(beef, 5) →;
Meat(pork, 7) →;

Fish(sole, 2) →;
Fish(tuna, 4) →;

Dessert(fruit, 2) →;
Dessert(icecream, 6) →.
```

The meaning of the first rule is: " provided the four conditions $i \geq 0$, $j \geq 0$, $k \geq 0$, $i+j+k \leq 10$ are satisfied, the triple h,m,d constitutes a light meal, if h is an hors-d'œuvre with calorific value i, if m is a main course with calorific value j and if d is a dessert with calorific value k ." The meaning of the last rule is: " Ice-cream is a dessert with calorific value 6 ".

Let us now give a precise definition of the set of admissible elements. The rules in the program are in fact rule schemas. Each rule (of the above form) stands for the set of *evaluated rules*

$$\sigma^*(t_0) \to \sigma^*(t_1)...\sigma^*(t_n)$$

obtained by considering all the solutions σ of S for which the values $\sigma^*(t_i)$ are defined. Each evaluated rule

$$a_0 \to a_1...a_n,$$

in which only elements a_i of the domain occur, can be interpreted in two ways:
(1) as a *closure property* of certain subsets E of the domain: if all of $a_1,...,a_n$ are members of the subset E, then a_0 is also is a member of E   (when $n = 0$, this property states that a_0 is a member of E),
(2) as a *rewrite rule* which, given a sequence of elements of the domain beginning with a_0, sanctions the replacement of this first element a_0 by the sequence $a_1...a_n$ (when $n = 0$, this is the same as deleting the first element a_0).

According to which of the two above interpretations is being considered, we formulate one or the other of the following definitions:

DEFINITION 1. The set of *admissible* elements is the smallest subset of the domain (in the sense of inclusion) which satisfies all the closure properties stemming from the program.

DEFINITION 2. The *admissible* elements are the members of the domain which (considered as unary sequences) can be deleted by applying rewrite rules stemming from the program a finite number of times.

In [8,11] we show that the smallest subset in the first definition does indeed exist and that the two definitions are equivalent. Let us re-examine the previous program example. Here are some samples of evaluated rules:

....
LightMeal(pâté,sole,fruit) →
        HorsDœuvre(pâté,6) MainCourse(sole,2) Dessert(fruit,2) ;
....
....
MainCourse(sole, 2) → Fish(sole,2) ;
....
....
HorsDœuvre(pâté,6) → ;
....
....
Fish(sole,2) → ;
....
....
Dessert(fruit,2) → ;
....

If we consider these rules to be closure properties of a subset of trees, we can successively conclude that the following three subsets are sets of admissible elements,

{HorsDœuvre(pâté,6), Fish(sole,2), Dessert(fruit,2)},
{MainCourse(sole,2)},
{LightMeal(pâté,sole,fruit)}

and therefore that the tree

LightMeal(pâté,sole,fruit)

is an admissible element. If we take these evaluated rules to be rewrite rules, the sequence constituted solely by the last tree can be deleted in the following rewrite steps

LightMeal(pâté,sole,fruit) →
HorsDœuvre(pâté,6) MainCourse(sole,2) Dessert(fruit,2) →
MainCourse(sole,2) Dessert(fruit,2) →

$$\text{Fish(sole,2) Dessert(fruit,2)} \rightarrow$$
$$\text{Dessert(fruit,2)} \rightarrow,$$

which indeed confirms that the above is an admissible element.

Execution of a program

We have now described the information implicit in a Prolog III like program, but we have not yet explained how such a program is executed. The aim of the program's execution is to solve the following problem: given a sequence of terms $t_1...t_n$ and a system S of constraints, find the values of the variables which transform all the terms t_i into admissible elements, while satisfying all the constraints in S. This problem is submitted to the machine by writing the *query*

$$t_1.... t_n , S ?$$

Two cases are of particular interest. (1) If the sequence $t_1...t_n$ is empty, then the query simply asks whether the system S is solvable and if so to solve it. We have already seen examples of such queries. (2) If the system S is empty (or absent) and the sequence of terms is reduced to one term only, the request can be summarized as: "What are the values of the variables which transform this term into an admissible element?". Thus using the preceding program example, the query

$$\text{LightMeal}(h, m, d)?$$

will enable us to obtain all the triples of values for h, m, and d which constitute a light meal. In this case, the replies will be the following simplified systems :

$$\{h = \text{radishes}, m = \text{beef}, d = \text{fruit}\},$$
$$\{h = \text{radishes}, m = \text{pork}, d = \text{fruit}\},$$
$$\{h = \text{radishes}, m = \text{sole}, d = \text{fruit}\},$$
$$\{h = \text{radishes}, m = \text{sole}, d = \text{icecream}\},$$
$$\{h = \text{radishes}, m = \text{tuna}, d = \text{fruit}\},$$
$$\{h = \text{pâté}, m = \text{sole}, d = \text{fruit}\}.$$

The method of computing these answers is explained by introducing an abstract machine. This is a non-deterministic machine whose state transitions are described by these three formulas :

$$(1) \quad (W, \ t_0 \ t_1...t_n \ , \ S),$$

$$(2) \qquad \quad s_0 \to s_1....s_m \ , \ R$$

$$(3) \quad (W, \ s_1...s_m \ t_1...t_n \ , \ S \cup R \cup \{t_0 = s_0\}).$$

Formula (1) represents the state of the machine at a given moment. W is a set of variables whose values we want to determine, $t_0 t_1...t_n$ is a sequence of terms which we are trying to delete and S is a system of constraints which has to be satisfied. Formula (2) represents the rule in the program which is used to change the state. If necessary, the variables of (2) are renamed, so that none of them are shared with (1). Formula (3) is the new state of the machine after the application of rule (2). The transition to this new state is possible only if the system of constraints in (3) possesses at least one solution σ with respect to which all the values $\sigma^*(s_i)$ and $\sigma^*(t_i)$ are defined.

In order to provide an answer to the query given above, the machine starts from the initial state

$$(W, \ t_0...t_n, \ S),$$

where W is the set of variables appearing in the query, and goes through all the states which can be reached by authorized transitions. Each time it arrives at a state containing the empty sequence of terms $\land$, it simplifies the system of constraints associated with it and presents it as an answer. This simplification can also be carried out on all the states it passes through.

Let us now reconsider our first program example, and apply this process to the query

$$\text{LightMeal}(h, m, d)?$$

The initial state of the machine is
$(\{h,m,d\}, \ \text{LightMeal}(h,m,d), \ \{\})$.
By applying the rule
$\text{LightMeal}(h', m', d') \to$
$\text{HorsD\oe uvre}(h', i) \ \text{MainCourse}(m', j) \ \text{Dessert}(d', k),$
$\{i \geq 0, j \geq 0, k \geq 0, i+j+k \leq 10\}$
we proceed to the state
$(\{h,p,d\}, \ \text{HorsD\oe uvre}(h',i) \ \text{MainCourse}(m',j) \ \text{Dessert}(d',k),$
$\{i \geq 0, j \geq 0, k \geq 0, i+j+k \leq 10, \text{LightMeal}(h,m,d) = \text{LightMeal}(h',m',d')\})$
which in turn simplifies to
$(\{h,p,d\}, \ \text{HorsD\oe uvre}(h',i) \ \text{MainCourse}(m',j) \ \text{Dessert}(d',k),$
$\{i \geq 0, j \geq 0, k \geq 0, i+j+k \leq 10, h=h', p=p', d=d'\}),$
and to
$(\{h,p,d\}, \ \text{HorsD\oe uvre}(h,i) \ \text{MainCourse}(m,j) \ \text{Dessert}(d,k),$
$\{i \geq 0, j \geq 0, k \geq 0, i+j+k \leq 10\}).$

By applying the rule
   HorsDœuvre(pâté, 6) →
and simplifying the result, we progress to the state
   ($\{h,p,d\}$, MainCourse(p,j) Dessert(d,k), $\{h$=pâté, $j{\geq}0$, $k{\geq}0$, $j{+}k{\leq}4\}$).
By applying the rule
   MainCourse(p', i) → Fish(p', i)
and simplifying the result, we proceed to the state
   ($\{h,m,d\}$, Fish(m',i) Dessert(d,k),
   $\{h$=pâté, $j{\geq}0$, $k{\geq}0$, $j{+}k{\leq}4$, $m{=}m'$, $j{=}i\}$).
which then simplifies to
   ($\{h,m,d\}$, Fish(m,j) Dessert(d,k), $\{h$=pâté, $j{\geq}0$, $k{\geq}0$, $j{+}k{\leq}4\}$).
By applying the rule
   Fish(sole, 2) →
we obtain
   ($\{h,m,d\}$, Dessert(d,k), $\{h$=pâté, m=sole, $k{\geq}0$, $k{\leq}2\}$).
Finally, by applying the rule
   Dessert(fruit, 2) →
we obtain
   ($\{h,m,d\}$, $\wedge$, $\{h$=pâté, m=sole, d=fruit$\}$).
We can conclude that the system

$$\{h = \text{pâté}, m = \text{sole}, d = \text{fruit}\}$$

constitutes one of the answers to the query.

To obtain the other answers, we proceed in the same way, but by using the other rules. In [11] we prove that this method is complete and correct. To be more exact, given the abstract machine M_P associated with a program P, we show that the following property holds.

PROPERTY. Let $\{t_1,...,t_n\}$ be a set of terms, S a system of constraints, and W the set of variables occurring in them. For any assignment σ to W, the following two propositions are equivalent :
(1) the assignment σ is a solution of S on W and each $\sigma^*(t_i)$ is an admissible element for P;
(2) starting from state (W, $\wedge$, S') the abstract machine M_P can reach a state of the form (W, $t_1...t_n$, S), where S' admits σ as solution on W.

It should be pointed out that there are a thousand ways of simplifying the states of the abstract machine and checking whether they contain solvable systems of constraints. So we should not always expect that the machine, which uses very general algorithms, arrives at the same simplifications as those that are shown above. In [11] we show that the only principle that simplifications must all conform to is that states of the abstract machine are transformed into equivalent states in this sense :

DEFINITION. Two states are *equivalent* if they have the form

$$(W, t_1...t_n , S) \text{ and } (W, t_1'...t_n', S'),$$

and if, by introducing n new variables $x_1,...,x_n$, the systems

$$S \cup \{x_1 = t_1,...,x_n = t_n\} \text{ and } S' \cup \{x_1 = t_1',...,x_n = t_n'\},$$

are equivalent on the subset of variables $W \cup \{x_1,...,x_n\}$.

TREATMENT OF NUMBERS

All that remains to be done is to illustrate the possibilities of Prolog III in connection with different program examples. We will consider one after the other the treatment of numbers, the treatment of Boolean values, the treatment of trees and lists and finally the treatment of integers.

Computing instalments

The first task is to calculate a series of instalments made to repay capital borrowed at a certain interest rate. We assume identical time periods between two instalments and an interest rate of 10% throughout. The admissible trees will be of the form :

$$\text{InstalmentsCapital}(x,c),$$

where x is the sequence of instalments necessary to repay the capital c with an interest rate of 10% between two instalments. The program itself is given by two rules :

```
InstalmentsCapital(<>, 0) →;
InstalmentsCapital(<i>•x, c) →
    InstalmentsCapital(x, (110/100)c−i);
```

The first rule expresses the fact that it is not necessary to pay instalments to repay zero capital. The second rule expresses the fact that the sequence of $n+1$ instalments to repay capital c consists of an instalment i and a sequence of n instalments to repay capital c increased by 10% interest, but the whole reduced by instalment i.

This program can be used in different ways. One of the most spectacular is to ask what value of i is required to repay $1000 given the sequence of instalments $<i,2i, 3i>$. All we need to do is to put the query

$$\text{InstalmentsCapital}(<i, 2i, 3i>, 1000) ?$$

to obtain the answer

$$\{i = 207 + 413/641\}.$$

Here is an abbreviated trace of how the computation proceeds. Starting from the initial state

$(\{i\}, \text{InstalmentsCapital}(<i,2i,3i>,1000), \{\})$.

and applying the rule

$\text{InstalmentsCapital}(<i'>\bullet x,c) \rightarrow \text{InstalmentsCapital}(x,(1+10/100)c-i')$

we progress to the state

$(\{i\}, \text{InstalmentsCapital}(x,(1+10/100)c-i'),$

$\{\text{InstalmentsCapital}(<i,2i,3i>,1000)=\text{InstalmentsCapital}(<i'>\bullet x,c)\})$,

which simplifies to

$(\{i\}, \text{InstalmentsCapital}(x,(11/10)c-i'), \{i'=i, x=<2i,3i>, c=1000\})$,

then to

$(\{i\}, \text{InstalmentsCapital}(<2i,3i>,1100-i), \{\})$.

The reader can verify that when the same rule is applied two more times, we obtain, after simplification, the states

$(\{i\}, \text{InstalmentsCapital}(<3i>,1210-(31/10)i), \{\})$,

$(\{i\}, \text{InstalmentsCapital}(<>,1331-(641/100)i), \{\})$.

By applying the rule

$\text{InstalmentsCapital}(<>,0) \rightarrow$

to the last state, we finally obtain

$(\{i\}, , \{1331-(641/100)i=0\}$

which simplifies to

$(\{i\}, , \{i=207+413/641\})$.

Here again the reader should be aware that the simplifications presented above are not necessarily those the machine will perform.

Computing scalar products

As an example of approximated multiplication, here is a small program which computes the scalar product $x_1 \times y_1 + ... + x_n \times y_n$ of two vectors $<x_1,...,x_n>$ and $<y_1,...,y_n>$.

```
ScalarProduct(<>, <>, 0) →;
ScalarProduct(<x>•X, <y>•Y, u+z) →
    ScalarProduct(X, Y, z),
    {u ≐ x×y};
```

The query

$$\text{ScalarProduct}(<1,1>, X, 12) \; \text{ScalarProduct}(X, <2,4>, 34) ?$$

produces the answer

$$\{X = <7,5>\}.$$

Computing the periodicity of a sequence

This problem was proposed in [5]. We consider the infinite sequence of real numbers defined by

$$x_{i+2} = |x_{i+1}| - x_i$$

where x_1 and x_2 are arbitrary numbers. Our aim is to show that this sequence is always periodic and that the period is 9, in other words, that the sequences

$$x_1, x_2, x_3, \ldots \quad \text{and} \quad x_{1+9}, x_{2+9}, x_{3+9}, \ldots$$

are always identical.

Each of these two sequences is completely determined if its first two elements are known. To show that the sequences are equal, it is therefore sufficient to show that in any sequence of eleven elements

$$x_1, x_2, x_3, \ldots, x_{10}, x_{11}$$

we have

$$x_1 = x_{10} \quad \text{and} \quad x_2 = x_{11}.$$

To begin with, here is the program that enumerates all the finite sequences $x_1, x_2, \ldots, x_n$ which respect the rule given above :

```
Sequence(<+y, +x>) →;
Sequence(<z,y,x>•s) →
    Sequence(<y,x>•s)
    AbsoluteValue(y, y'), {z = y'−x};

AbsoluteValue(y, y) →, {y ≥ 0};
AbsoluteValue(y, −y) →, {y < 0};
```

The + signs in the first rule constrain x and y to denote numbers. It will be observed that the sequences are enumerated from left to right, that is, trees of the form Sequence(s) are admissible only if s has the form $<x_n, \ldots, x_2, x_1>$. If we run this program by asking

$$\text{Sequence}(s), \{|s| = 11, s \doteq w \bullet v \bullet u, |u| = 2, |w| = 2, u \neq w\} \ ?$$

execution ends without providing an answer. From this we deduce that there is no sequence of the form $x_1, x_2, \ldots, x_{10}, x_{11}$ such that the subsequences x_1, x_2 and x_{10}, x_{11} (denoted by u and v) are different, and therefore that in any sequence $x_1, x_2, \ldots, x_{10}, x_{11}$ we have indeed $x_1 = x_{10}$ and $x_2 = x_{11}$.

Here is a last example which highlights the numerical part of Prolog III. Given an integer n, we want to know whether it is possible to have n squares of different sizes which can be assembled to form a rectangle. If this is possible, we would in addition like to determine the sizes of these squares and of the rectangle thus formed. For example, here are two solutions to this problem, for $n=9$.

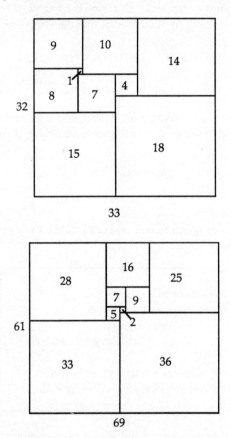

We will use a to denote the ratio between the length of the longest side of the constructed rectangle, and the length of its shortest side. Obviously, we can suppose that the length of the shortest side is 1, and therefore that the length of the longest side is a. Thus, we have to fill a rectangle having the size $1 \times a$ with n squares, all of them different. With reference to the diagram below, the basis of the filling algorithm will consist of
(1) placing a square in the lower left-hand corner of the rectangle,
(2) filling zone A with squares,
(3) filling zone B with squares.
Provided zones A and B are not empty, they will be filled recursively in the same way: placing a square in the lower left-hand corner and filling two subzones.

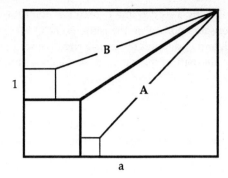

a

The zones and subzones are separated by jagged lines in the shape of steps, joining the upper right corner of the squares and the upper right corner of the rectangle. These jagged lines never go downwards, and if several can be plotted to go from one point to another, the lowest one is the one which is considered. Here are for example all the separation lines corresponding to the first solution of the problem for n = 9 :

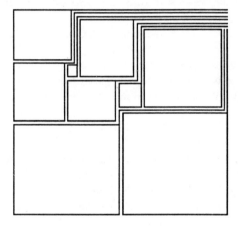

To be more precise, a zone or subzone has the form given in the left diagram below, whereas the entire rectangle is itself identified with the particular zone drawn on the right.

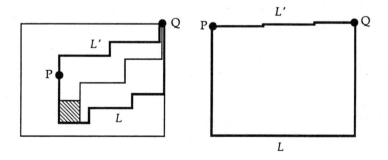

The zone is delimited below by a jagged line L joining a point P to a point Q, and above by a jagged line L' joining the same point P to the same point Q. Point P is placed anywhere in the rectangle to be filled, and Q denotes the upper right corner of the rectangle. These jagged lines are represented by alternating sequences of vertical and horizontal segments

$$v_0, h_1, v_1, \dots, h_n, v_n,$$

where v_i denotes the length of a vertical segment, and h_i the length of a horizontal segment. The h_i's are always strictly positive. The v_i's are either zero, either positive to denote ascending segments, or negative to denote descending segments. The v_i's of the upper lines are never negative, and if a zone is not empty, only the first vertical segment v_0 in its lower line is negative.

If theses conventions are applied to the entire rectangle (right diagram above), the lower line L can be represented by the sequence -1,a,1 and the upper line L' by a sequence having the form $0,h_1,0,\dots,h_n,0$, where $h_1+\dots+h_n = a$, and the h_i's are positive.

The heart of the program consists in admitting trees of the form

FilledZone(L, L', C, C')

only if the zone delimited below by L can be filled with squares and can be bounded above by L'. The squares are to be taken from the beginning of the list C, and C' has to be the list of squares which remain. We also need to introduce trees of the form

PlacedSquare(b, L, L')

which are admitted only if it is possible to place a square of size bxb at the very beginning of line L and if L' is the line making up the right vertical side of the square continued by the right part of L (see diagram below). In fact L denotes the lower line of a zone from which the first vertical segment has been removed. The diagram below shows the three cases that can occur and which will show up in three rules. Either the square overlaps the first step, which in fact was a pseudostep of height zero, or the square fits against the first step, or the square is not big enough to reach the first step.

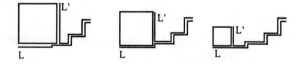

The program itself is constituted by the following ten rules:

FilledRectangle(*a*, *C*) →
   DistinctSquares(*C*)
   FilledZone(<–1,*a*,1>, *L*, *C*, <>),
   {*a* ≥ 1};

DistinctSquares(<>) →;
DistinctSquares(<*b*>•*C*) →
   DistinctSquares(*C*)
   OutOf(*b*, *C*),
   {*b* > 0};

OutOf(*b*, <>) →;
OutOf(*b*, <*b'*>•*C*) →
   OutOf(*b*, *C*),
   {*b* ≠ *b'*};

FilledZone(<*v*>•*L*, <*v*>•*L*, *C*, *C*) →,
   {*v* ≥ 0};
FilledZone(<*v*>•*L*, *L'''*, <*b*>•*C*, *C''*) →
   PlacedSquare(*b*, *L*, *L'*)
   FilledZone(*L'*, *L''*, *C*, *C'*)
   FilledZone(<*v*+*b*,*b*>•*L''*, *L'''*, *C'*, *C''*),
   {*v* < 0};

PlacedSquare(*b*, <*h*,0,*h'*>•*L*, *L'*) →
   PlacedSquare(*b*, <*h*+*h'*>•*L*, *L'*),
   {*b* > *h*};
PlacedSquare(*b*, <*h*,*v*>•*L*, <–*b*+*v*>•*L*) →,
   {*b* = *h*};
PlacedSquare(*b*, <*h*>•*L*, <–*b*,*h*–*b*>•*L*) →,
   {*b* < *h*};

The call to the program is made with the query

$$\text{FilledRectangle}(a, C),\ \{|C| = n\}\ ?$$

where *n*, the only known parameter, is the number of squares of different sizes that are to fill the rectangle. The program computes the possible size 1×*a* of the rectangle (*a*≥1) and the list C of the sizes of each of the *n* squares. The computation begins by executing the first rule, which at the same time constrains *a* to be greater than or equal to 1, creates *n* different squares (of unknown size) and starts filling the zone constituted by the entire rectangle. Even if the line L constituting the upper limit of this zone is unknown at the beginning, given that this line must join - without itself descending - two points at the same height, this line will necessarily be a horizontal line (represented by steps of height zero). If we ask the query

$$\text{FilledRectangle}(a, C), \{|C| = 9\} \text{ ?}$$

we obtain 8 answers. The first two

$\{a = 33/32, C = <15/32, 9/16, 1/4, 7/32, 1/8, 7/16, 1/32, 5/16, 9/32>\}$,

$\{a = 69/61, C = <33/61, 36/61, 28/61, 5/61, 2/61, 9/61, 25/61, 7/61, 16/61>\}$.

correspond to the two solutions we have drawn earlier. The other 6 answers describe solutions which are symmetrical to these two. In order to locate the positions of the various squares in the rectangle we can proceed as follows. One fills the rectangle using one after the other all the squares of the list C in their order of appearance. At each stage one considers all the free corners having the same orientation as the lower left corner of the rectangle and one chooses the rightmost one to place the square.

There is a vast literature concerning the problem that we have just dealt with. Let us mention two important results. It has been shown in [25] that for any rational number $a \geq 1$ there always exists an integer n such that the rectangle of size $1 \times a$ can be filled with n distinct squares. For the case of $a = 1$, that is when the rectangle to be filled is a square, it has been shown in [14] that the smallest possible n is $n = 21$.

TREATMENT OF BOOLEAN VALUES

Computing faults

In this example we are interested in detecting the defective components in an adder which calculates the binary sum of three bits x_1, x_2, x_3 in the form of a binary number given in two bits $y_1 y_2$. As we can see below, the circuit proposed in [16] is made up of 5 components numbered from 1 to 5: two *and* gates (marked And), one *or* gate (marked Or) and two *exclusive or* gates (marked Xor). We have also used three variables u_1, u_2, u_3 to represent the output from gates 1, 2 and 4.

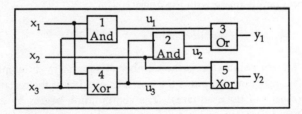

We introduce 5 more Boolean variables d_i to express by $d_i = 1'$ that "gate number i is defective". If we adopt the hypothesis that at most one of the five components has a defect, the program connecting the values x_i, y_i and d_i is :

```
Circuit(<x1,x2,x3>, <y1,y2>, <d1,d2,d3,d4,d5>) →
     AtMostOne(<d1,d2,d3,d4,d5>),
     {¬d1 ⇒ (u1 ≡ x1∧x3),
     ¬d2 ⇒ (u2 ≡ x2∧u3),
     ¬d3 ⇒ (y1 ≡ u1∨u2),
     ¬d4 ⇒ (u3 ≡ ¬(x1≡x3)),
     ¬d5 ⇒ (y2 ≡ ¬(x2≡u3))};


AtMostOne(D) →
     OrInAtMostOne(D, d);


OrInAtMostOne(<>, 0') →;
OrInAtMostOne(<d>•D, d∨e) →
     OrInAtMostOne(D, e),
     {d∧e = 0'};
```

In this program the admissible trees of the form

$$AtMostOne(D)$$

are those in which D is a list of Boolean elements containing at most one 1'. The admissible trees of the form

$$OrInAtMostOne(D, d)$$

are those in which D is a list of Boolean elements containing at most one 1' and where d is the disjunction of these elements.

If the state of the circuit leads us to write the query

$$Circuit(<1', 1', 0'>, <0', 1'>, <d1, d2, d3, d4, d5>)?$$

the diagnosis will be that component number 4 is defective :

$$\{ d1= 0', \ d2 = 0', \ d3 = 0', d4 = 1', d5 = 0' \}.$$

If the state of the circuit leads us to write the query

$$Circuit(<1', 0', 1'>, <0', 0'>, <d1, d2, d3, d4, d5>) ?$$

the diagnosis will then be that either component number 1 or component number 3 is the defective one:

$$\{ d1∨d3 = 1', d1∧d3 = 0', d2 = 0', \ d4 = 0', d5 = 0'\}.$$

Computing inferences

We now consider the 18 sentences of a puzzle due to Lewis Carroll [7], which we give below. Questions of the following type are to be answered : "what connection is there between being clear-headed, being popular and being fit to be a Member of Parliament?" or "what connection is there between being able to keep a secret, being fit to be a Member of Parliament and being worth one's weight in gold ?".

1. Any one, fit to be an M.P., who is not always speaking, is a public benefactor.

2. Clear-headed people, who express themselves well, have a good education.

3. A woman, who deserves praise, is one who can keep a secret.

4. People, who benefit the public, but do not use their influence for good purpose, are not fit to go into Parliament.

5. People, who are worth their weight in gold and who deserve praise, are always unassuming.

6. Public benefactors, who use their influence for good objects, deserve praise.

7. People, who are unpopular and not worth their weight in gold, never can keep a secret.

8. People, who can talk for ever and are fit to be Members of Parliament, deserve praise.

9. Any one, who can keep a secret and who is unassuming, is a never-to-be-forgotten public benefactor.

10. A woman, who benefits the public, is always popular.

11. People, who are worth their weight in gold, who never leave off talking, and whom it is impossible to forget, are just the people whose photographs are in all the shop-windows.

12. An ill-educated woman, who is not clear-headed, is not fit to go to Parliament.

13. Any, one, who can keep a secret and is not for ever talking, is sure to be unpopular.

14. A clear-headed person, who has influence and uses it for good objects, is a public benefactor.

15. A public benefactor, who is unassuming, is not the sort of person whose photograph is in every shop-window.

16. People, who can keep a secret and who who use their influence for good purposes, are worth their weight in gold.

17. A person, who has no power of expression and who cannot influence others, is certainly not a woman.

18. People, who are popular and worthy of praise, either are public benefactors ore else are unassuming.

Each of these 18 statements is formed from basic propositions and logical connectives. To each basic proposition corresponds a name, in the form of a character string, and a logical value represented by a Boolean variable. The information contained in the 18 statements can then be expressed in a single rule formed by a large head term, an empty body, and a sizeable constraint part :

PossibleCase(<
<a,"clear-headed">,
<b,"well-educated">,
<c,"constantly talking">,
<d,"using one's influence for good objects">,
<e,"exhibited in shop-windows">,
<f,"fit to be a Member of Parliament">,
<g,"public benefactors">,
<h,"deserving praise">,
<i,"popular">,
<j,"unassuming">,
<k,"women">,
<l,"never-to-be-forgotten">,
<m,"influential">,
<n,"able to keep a secret">,
<o,"expressing oneself well">,
<p,"worth one's weight in gold">>)   →,
$$\{(f \wedge \neg c) \Rightarrow g,$$
$$(a \wedge o) \Rightarrow b,$$
$$(k \wedge h) \Rightarrow n,$$
$$(g \wedge \neg d) \Rightarrow \neg f,$$
$$(p \wedge h) \Rightarrow j,$$
$$(g \wedge d) \Rightarrow h,$$
$$(\neg i \wedge \neg p) \Rightarrow \neg n,$$
$$(c \wedge f) \Rightarrow h,$$
$$(n \wedge j) \Rightarrow (g \wedge l),$$
$$(k \wedge g) \Rightarrow i,$$
$$(p \wedge c \wedge l) \Rightarrow e,$$
$$(k \wedge \neg a \wedge \neg b) \Rightarrow \neg f,$$
$$(n \wedge \neg c) \Rightarrow \neg i,$$
$$(a \wedge m \wedge d) \Rightarrow g,$$
$$(g \wedge j) \Rightarrow \neg e,$$
$$(n \wedge d) \Rightarrow p,$$
$$(\neg o \wedge \neg m) \Rightarrow \neg k,$$
$$(i \wedge h) \Rightarrow (g \vee j)\};$$

To be able to deal with subcases, we introduce :

```
PossibleSubCase(x) →
    PossibleCase(y)
    SubSet(x, y);

SubSet(<>, y) →;
SubSet(<e>•x, y) →
    ElementOf(e, y)
    SubSet(x, y);

ElementOf(e, <e>•y) →;
ElementOf(e, <e'>•y) →
    ElementOf(e, y), {e ≠ e' };
```

In order to compute the connection which exists between "clear-headed", "popular"and "fit to be a Member of Parliament" it suffices to write the query

    PossibleSubCase(<
        $<p,$"clear-headed">,
        $<q,$"popular">,
        $<r,$"fit to be a Member of Parliament">>)?

The answer is the set of constraints

    $\{p : \text{bool}, q : \text{bool}, r : \text{bool}\}$,

which means that there is no connection between "clear-headed", "popular" and "fit to be a Member of Parliament".

To compute the connection which exists between "able to keep a secret", "fit to be a Member of Parliament" and "worth one's weight in gold" it suffices to write the query

    PossibleSubCase(<
        $<p,$"able to keep a secret">,
        $<q,$"fit to be a Member of Parliament">,
        $<r,$"worth one's weight in gold">>) ?

The answer is

    $\{p \wedge q \Rightarrow r\}$,

which means that persons who can keep a secret and are fit to be a Member of Parliament are worth their weight in gold.

In fact in these two examples of program execution we have assumed that Prolog III yields as answer very simplified solved systems, in particular, systems not containing superfluous Boolean variables. If this head not been the case, to show (as opposed to find) that persons who can keep a secret and are fit to be a Member of Parliament are worth their weight in gold, we would have had to pose the query

    PossibleSubCase(<
        <p,"able to keep a secret">,
        <q,"fit to be a Member of Parliament">,
        <r,"worth one's weight in gold">>),
    $\{x = (p \land q \supset r)\}$?

and obtain a response of the form $\{x = 1', ... \}$ or obtain no answer to the query

    PossibleSubCase(<
        <p,"able to keep a secret">,
        <q,"fit to be a Member of Parliament">,
        <r,"worth one's weight in gold">>),
    $\{(p \land q \supset r) = 0'\}$?

TREATMENT OF TREES AND LISTS

Computing the leaves of a tree

Here is first of all an example where we access labels and daughters of a tree by the operation []. We want to calculate the list of the leaves of a finite tree without taking into account the leaves labeled $<>^{\alpha}$. Here is the program :

```
Leaves(e[u], <e>) →,
    {u = <>};
Leaves(e[u], x) →
    Leaves(u, x),
    {u ≠ <>};
Leaves(<>, <>) →;
Leaves(<a>•u, z) →
    Leaves(a, x)
    Leaves(u, y),
    {z ≐ x•y};
```

Trees of the form

$$Leaves(a, x)$$

are admissible only if x is the list of leaves of the finite tree a (not including the leaves labeled $<>^{\alpha)}$. The query

$$\text{Leaves(height("Max",<180/100,meters>,1'), } x)?$$

produces the answer

$$\{x = <'M', \text{'a', 'x'}, 9/5, \text{meters}, 1'>\}.$$

Computing decimal integers

Our second example shows how approximated concatenation can be used to access the last element of a list. We want to transform a sequence of digits into the integer it represents. Here is the program without comments :

```
Value(<>, 0) → ;
Value(y, 10m+n) →
    Value(x, m),
    {y ≐ x•<n>};
```

As a reply to the query

$$\text{Value(<1,9,9,0>, } x) ?$$

we obtain

$$\{x = 1990\}.$$

Computing the reverse of lists

If one knows how to access the first and the last element of a list it must be possible to write an elegant program computing the reverse of a list. Here is the one I propose :

Reverse$(x, y) \rightarrow$
    Palindrome(u),
    $\{u \doteq x \cdot y, |x| = |y|\}$;

Palindrome$(<>) \rightarrow$;
Palindrome$(v) \rightarrow$
    Palindrome(u),
    $\{v \doteq <a> \cdot u \cdot <a>\}$;

Each of the two queries

$$\text{Reverse}(<1,2,3,4,5>, x) ?$$
$$\text{Reverse}(x, <1,2,3,4,5>) ?$$

produces the same answer

$$\{x = <5,4,3,2,1>\}.$$

For the query

$$\text{Reverse}(x, y) \ \text{Reverse}(y, z), \{x \neq z, |x| = 10\} ?$$

we get no answer at all, which confirms that reversing a list twice yields the initial list.

Context-free recognizer

The treatment of concatenation provides a systematic and natural means of relating "context-free" grammar rules with Prolog III rules, thus constructing a recognizer. Let us for example consider the grammar

$$\{S \rightarrow AX, \ A \rightarrow \wedge, \ A \rightarrow aA, \ X \rightarrow \wedge, \ X \rightarrow aXb\}$$

which defines the language consisting of sequences of symbols of the form $a^m b^n$ with $m \geq n$. The following program corresponds to the grammar :

$$\begin{aligned}
&\text{Sform}(u) \to \\
&\quad \text{Aform}(v) \\
&\quad \text{Xform}(w), \\
&\quad \{u \doteq v \bullet w\}; \\[4pt]
&\text{Aform}(u) \to \\
&\quad \{u = <>\}; \\[4pt]
&\text{Aform}(u) \to \\
&\quad \text{Aform}(v), \\
&\quad \{u = \text{"a"} \bullet v\}; \\[4pt]
&\text{Xform}(u) \to \\
&\quad \{u = <>\}; \\[4pt]
&\text{Xform}(u) \to \\
&\quad \text{Xform}(v), \\
&\quad \{u \doteq \text{"a"} \bullet v \bullet \text{"b"}\};
\end{aligned}$$

The query

$$\text{Sform("aaabb")} ?$$

produces the answer

$$\{\}$$

which signifies that the string "aaabb" belongs to the language, whereas the query
$$\text{Sform("aaabbbb")} ?$$

produces no response, which means that the string "aaabbbb" does not belong to the language.

TREATMENT OF INTEGERS

The algorithms used for solving constraints on integers are complex and quite often inefficient. It is for this reason that the structure underlying Prolog III does not contain a relation restricting a number to be only an integer. We have however considered a way of enumerating integers satisfying the set of current constraints.

Enumeration of integers

The Prolog III abstract machine is modified in such a way as to behave as if the following infinite set of rules

$$enum(0) \rightarrow;$$
$$enum(-1) \rightarrow;$$
$$enum(1) \rightarrow;$$
$$enum(-2) \rightarrow;$$
$$enum(2) \rightarrow;$$
$$\dots\dots\dots\dots$$

had been added to every program. Moreover the abstract machine is implemented in such a way as to guarantee that the search for applicable rules takes a finite among of time whenever this set is itself finite. In connection with the definition of the abstract machine this can be regarded as adding all the transitions of the form

$$(W, t_0 \; t_1 ... t_m , S) \quad \rightarrow \quad (W, t_1 ... t_m , S \cup \{t_0 = enum(n)\}),$$

where **n** is an integer such that the system $S \cup \{p_0 = enum(n)\}$ admits at least one solution in which the values of the t_i 's are all defined.

For example, if in the current state of the abstract machine the first term to be deleted is « enum(x) » and if the system S of constraints is equivalent on $\{x\}$ to $\{3/4 \leq x, x \leq 3+1/4\}$, then they will be two transitions : one to a state with a system equivalent to $S \cup \{x=1\}$, the other to a state with a system equivalent to $S \cup \{x=2\}$.

Let us add in this connection that if S is a system forcing the variable x to represent a number, then, in the most complex case, the system S is equivalent on $\{x\}$ to a system of the form

$$\{x \geq a_0, x \neq a_1, ... , x \neq a_n, x \leq a_{n+1}\},$$

where the a_i 's are rational numbers.

A problem, taken from one of the many books of M. Gardner [15], illustrates nicely the enumeration of integers. The problem goes like this. In times when prices of farm animals were much lower than today, a farmer spent \$100 to buy 100 animals of three different kinds, cows, pigs and sheep. Each cow cost \$10, each pig \$3 and each sheep 50 cents. Assuming that he bought at least one cow, one pig and one sheep, how many of each animal did the farmer buy ?

Let x, y and z be the number of cows, pigs and sheep that the farmer bought. The query

$$enum(x) \; enum(y) \; enum(z),$$
$$\{x+y+z = 100, 10x+3y+z/2 = 100, x \geq 1, y \geq 1, z \geq 1\} \; ?$$

produces the answer

$$\{x = 5, y = 1, z = 94\}.$$

This problem reminds us of a problem mentioned at the beginning of this paper. Find the number x of pigeons and the number y of rabbits such that together there is a total of 12 heads and 34 legs. It was solved by putting the query

$$\{x+y = 12, 2x+4y = 34\} \ ?$$

But, given that, a priori, we have no guarantee that the solutions of this system are non-negative and integer numbers, it is more appropriate to put the query

$$\text{enum}(x) \ \text{enum}(y), \ \{x+y = 12, 2x+4y = 34, x \geq 0, y \geq 0\} \ ?$$

which produces the same answer

$$\{x = 7, y = 5\}.$$

Cripto-arithmetic

Here is another problem that illustrates the enumeration of integers. We are asked to solve a classical cripto-arithmetic puzzle : assign the ten digits $0,1,2,3,4,5,6,7,8,9$ to the ten letters D,G,R,O,E,N,B,A,L,T in such a way that the addition $DONALD + GERALD = ROBERT$ holds. We deterministically install the maximal number of constraints on the reals and use the non-determinism to enumerate all the integers which are to satisfy these constraints. Here is the program without any comments :

```
Solution(i, j, i+j) →
        Value(<D,O,N,A,L,D>, i)
        Value(<G,E,R,A,L,D>, j)
        Value(<R,O,B,E,R,T>, i+j)
        DifferentAndBetween09(x)
        Integers(x),
        {<D,G,R,E,N,B,A,L,T,O> = x,
        D ≠ 0, G ≠ 0, R ≠ 0};

Value(<>, 0)→ ;
Value(y, 10i+j) →
        Value(x, i), {y ≐ x•<j>};

DifferentAndBetween09(<>) →;
DifferentAndBetween09(<i>•x) →
        OutOf(i, x)
        DifferentAndBetween09(x),
        {0 ≤ i, i ≤ 9};

OutOf(i, <>) →;
OutOf(i, <j>•x) →
        OutOf(i, x), {i ≠ j};

Integers(<>) →;
Integers(<i>•x) →
        enum(i) Integers(x);
```

The answer to the query

$$Solution(i, j, k) ?$$

is

$$\{i = 526485, j = 197485, k = 723970\}.$$

Self-referential puzzle

The last example is a typical combinatorial problem that is given a natural solution by enumeration of integers in a involving approximated concatenation and multiplication. Given a positive integer n, we are asked to find n integers $x_1, .., x_n$ such that the following property holds :

"In the sentence that I am presently uttering, the number 1 occurs x_1 times, the number 2 occurs x_2 times, ... , the number n occurs x_n times".

We proceeds as if one were using true (and not approximated) concatenation and one writes the program whose admissible trees are of the form

$$\text{Counting}(<x_1,...,x_m>, <y_1+1,...,y_n+1>),$$

each x_i being an integer between 0 and m, each y_i being the number of occurrences of the integer i in the list $<x_1,...,x_m>$. This is the program :

> Counting(<>, Y) →,
>   {<1>•Y = Y•<1>};
> Counting(<x>•X, U•<y+1>•V) →
>   Counting(X, U•<y>•V),
>   {|U| = x–1};

Here the constraint $\{<1>•Y = Y•<1>\}$ is an elegant way of forcing Y to be a list of 1's. If everything were perfect, it would suffice to ask the query "Counting(X, X), $\{|X| = n\}$" to obtain the list of the desired n integers. Prolog III not being perfect, we have to substitute approximated concatenation for true concatenations. We must therefore complete the program with an enumeration of the integers $x_1, .., x_n$ that we are looking for. All the lists are thus constrained to be of integer length, that is to say, to be true lists and as a result all the approximated concatenations become true concatenations. In order to reduce the enumeration of integers, we introduce two properties: The first property is

$$x_1+ ... +x_n = 2n,$$

which expresses that the total number of occurrences of numbers in the sentences is both $x_1+...+x_n$ and $2n$. The second is

$$0x_1+1x_2+ ... +(n-1)x_n = n(n+1)/2,$$

which expresses that the sum of numbers which appear in the sentences is both $1x_1+2x_2+...+nx_n$ and $x_1+... +x_n + 1+...+n$. From all these considerations the following final program results :

```
Solution(X) →
     Sum(X, 2n)
     WeightedSum(X, m)
     Counting(X, X)
     Integers(X),
     {n = |X|, m ≐ n×̇(n+1)/2};

Sum(<>, 0) →;
Sum(<x>•X, x+y) →
     Sum(X, y);

WeightedSum(<>, 0) →;
WeightedSum(X•<x>, z+y) →
     WeightedSum(X, y),
     {z ≐ |X|×̇x};

Counting(<>, Y) →,
     {<1>•Y ≐ Y•<1>};
Counting(<x>•X, Y') →
     Counting(X, Y),
     {Y' ≐ U•<y+1>•V,
     Y ≐ U•<y>•V,
     |U| = x−1};

Integers(<>) →;
Integers(<x>•X) →
     Integers(X)
     enum(x);
```

Assigning successively to n the values 1,2, ... , 20 and asking the query

$$\text{Solution}(X), \{|X| = n\} \ ?$$

we obtain as answers

{X = <3,1,3,1>},
{X = <2,3,2,1>},
{X = <3,2,3,1,1>},
{X = <4,3,2,2,1,1,1>},
{X = <5,3,2,1,2,1,1,1>},
{X = <6,3,2,1,1,2,1,1,1>},
{X = <7,3,2,1,1,1,2,1,1,1>},
{X = <8,3,2,1,1,1,12,1,1,1>},
....
{X = <16,3,2,1,1,1,1,1,1,1,1,1,1,1,2,1,1,1>},

$$\{X = <17,3,2,1,1,1,1,1,1,1,1,1,1,1,1,1,2,1,1,1>\}.$$

The regularity in the answer gives rise to the idea of proving that for $n \geq 7$ there always exists a solution of the form

$$x_1, \dots , x_n \;\; = \;\; n{-}3, 3, 2, 1, \dots , 1, 2, 1, 1, 1.$$

PRACTICAL REALIZATION

Prolog III is of course more than an intellectual exercise. A prototype of a Prolog III interpreter has been running in our laboratory since the end of 1987. A commercial version based on this prototype is now being distributed by the company PrologIA at Marseilles (Prolog III version 1). This product incorporates the functionalities described in the present paper as well as facilities calculating maximum and minimum values of numerical expressions. We have been able to use it to test our examples and to establish the following benchmarks (on a Mac II, first model).

Light meals	4 sec
Instalments, $n = 3$	2 sec
Instalments, $n = 50$	6 sec
Instalments, $n = 100$	23 sec
Periodic sequence	3 sec
Squares, $n = 9$	13 min 15 sec
Squares, $n = 9$, 1st solution	1 min 21 sec
Squares, $n = 10$, 1st solution	6 min 36 sec
Squares, $n = 11$, 1st solution	1 min 38 sec
Squares, $n = 12$, 1st solution	5 min 02 sec
Squares, $n = 13$, 1st solution	4 min 17 sec
Squares, $n = 14$, 1st solution	13 min 05 sec
Squares, $n = 15$, 1st solution	11 min 29 sec
Faults detection, 2nd query	3 sec
Lewis Carrol, 2nd query	3 sec
Donald+Gerald...	68 sec
Self-referential-puzzle, $n = 4$	3 sec
Self-referential-puzzle, $n = 5$	4 sec
Self-referential-puzzle, $n = 10$	11 sec
Self-referential-puzzle, $n = 15$	36 sec
Self-referential-puzzle, $n = 20$	1 min 54 sec
Self-referential-puzzle, $n = 25$	5 min 51 sec
Self-referential-puzzle, $n = 30$	17 min 55 sec

All the above figures, except when stated otherwise, are the execution times of complete programs including the backtracking, input of queries and output of answers. The instalment calculation consist in computing a sequence of instalments $i, 2i, 3i, \ldots, ni$ needed to reimburse a capital of 1000. In order to do justice to these results one must take into account the fact that all the calculations are carried out in infinite precision. In the instalment example with $n=100$ a simplified fraction with a numerator and a denominator with more than 100 digits is produced !

We finish this paper with information on the implementation of Prolog III. The kernel of the Prolog III interpreter consists of a two-stack machine which explores the search space of the abstract machine via backtracking. These two stacks are filled and emptied simultaneously. In the first stack one stores the structures representing the states through which one passes. In the second stack one keeps track of all the modifications made on the first stack by address-value pairs in order to make the needed restorations upon backtracking. A general system of garbage collection [23] is able to detect those structures that have become inaccessible and to recuperate the space they occupy by compacting the two stacks. During this compaction the topography of the stacks is completely retained. The kernel of the interpreter also contains the central part of the solving algorithms for the = and ≠ constraints. These algorithms are essentially an extension of those already used in Prolog II and described in [10]. The extension concerns the treatment of list concatenation and the treatment of linear numerical equations containing at least one variable not restricted to represent a non-negative number. A general mechanism for the delaying of constraints, which is used to implement approximated multiplication and concatenation, is also provided in the kernel. Two submodules are called upon by the interpreter, one for the treatment of Boolean algebra, the other for the remaining numerical part.

The Boolean algebra module works with clausal forms. The algorithms used [2] are an incremental version of those developed by P. Siegel [24], which are themselves based on SL-resolution [20]. On the one hand they determine if a set of Boolean constraint is solvable and on the other they simplify these constraints into a set of constraints containing only a minimal subset of variables. Related experiments have been performed with an algorithm based on model enumeration [21]. Although significant improvement has been achieved as far as solvability tests are concerned a large part of these ameliorations is lost when it comes to simplifying the constraints on output. Let us mention that W. Büttner and H. Simonis approach the incremental solving of Boolean constraints with quite different algorithms [6].

The numerical module treats linear equations the variables of which are constrained to represent non-negative numbers. (These variables x are introduced to replace constraints of the form $p \geq 0$ by the constraints $x = p$ and $x \geq 0$). The module consists essentially in an incremental implementation of G. Dantzig's simplex algorithm [12]. The choice of pivots follows a method proposed in M. Balinski et R. Gomory [1] which, like the well-known method of R. Bland [3], avoids cycles. The simplex algorithm is used both to verify if the numerical constraints have solutions and to detect those variables having only one possible value. This allows to simplify the constraints by detecting the hidden equations in the original constraints. For example the hidden equation $x = y$ will be detected in $\{x \geq y, y \geq x\}$. The module also contains various subprograms needed for addition and multiplication operations in infinite precision, that is to say, on fractions whose numerators and denominators are unbounded integers. Unfortunately we have not included algorithms for the systematic elimination of useless numerical variables in the solved systems of constraints. Let us mention in this connection the work of J-L. Imbert [17].

ACKNOWLEDGEMENTS

The prototype interpreter was built in cooperation between our laboratory (GIA) and the company PrologIA. Substantial financial support was obtained on the one hand from the Centre National d'Etudes des Télécommunications (contract 86 1B 027) and on the other from the CEE within the framework of the Esprit project P1106 "Further development of Prolog and its Validation by KBS in Technical Areas". Additional support was granted by the Commissariat à l'Energie Atomique in connection with the Association Méditérannéenne pour le Développement de l'IA, by the DEC company in connection with an External Research Grant and by the Ministère de la Recherche et de l'Enseignement Supérieur within the two "Programmes de Recherches Concertés", "Génie Logiciel" and "Intelligence Artificielle". Finally the most recent work on approximated multiplication and concatenation has been supported by the CEE Basic Research initiative in the context of the "Computational Logic" project.

I thank the entire research team which has been working on the Prolog III interpreter : Jean-Marc Boï and Frédéric Benhamou for the Boolean algebra module, Pascal Bouvier for the supervisor, Michel Henrion for the numerical module, Touraïvane for the kernel of the interpreter and for his work on approximated multiplication and concatenation. I also thank Jacques Cohen of Brandeis University whose strong interest has been responsible for my writing this paper and Franz Guenthner of the University of Tübingen who helped in the preparation of the final version of this paper. Finally I thank Rüdiger Loos of the University of Tübingen who pointed my attention to two particularly interesting numerical problems : the periodical sequence and the filling of a rectangle by squares.

REFERENCES

1. BALINSKI M. L. and R. E. GOMORY, A mutual primal-dual simplex method. In *Recent Advances in Mathematical Programming*, R.L. Graves and P. Wolfe, Eds. McGraw-Hill, New York, 1963, pp. 17-26.

2. BENHAMOU F. and J-M. BOI, *Le traitement des contraintes Booléennes dans Prolog III*, Thèses de Doctorat, GIA, Faculté des Sciences de Luminy, Université Aix-Marseille II, Novembre 1988.

3. BLAND R. G., New finite pivoting for the simplex method, *Mathematics of Operations Research*, Vol. 2, No. 2, May 1977.

4. BOOLE G., *The Laws of Thought*, Dover Publication Inc., New York, 1958.

5. BROWN M., Problem proposed in : *The American Mathematical Monthly*, vol. 90, no. 8, pp. 569, 1983.

6. BÜTTNER W. and H. SIMONIS, Embedding Boolean Expressions into Logic Programming, *Journal of Symbolic Computation*, 4, 1987.

7. CARROLL L., *Symbolic Logic and the Game of Logic*, New York, Dover, 1958.

8. COLMERAUER A., Theoretical model of Prolog II, *Logic programming and its application*, ed. by M. Van Caneghem and D. Warren, Ablex Publishing Corporation, pp. 3-31, 1986.

9. COLMERAUER A., Prolog in 10 figures, *Communications of the ACM*, Volume 28, Number 12, pp. 1296-1310, December 1985.

10. COLMERAUER A., Equations and Inequations on Finite and Infinite Trees, Invited lecture, *Proceedings of the International Conference on Fifth Generation Computer Systems*, Tokyo, pp. 85-99, November 1984.

11. COLMERAUER A., *Final Specifications for Prolog III*, Esprit P11O6 report : Further development of Prolog and its Validation by KBS in Technical Areas, February, 1988.

12. DANTZIG G. B., *Linear Programming and Extensions*, Princeton University Press, 1963.

13. DINCBAS M. and AL., The Constraint Logic Programming CHIP, *Proceedings of the International Conference on Fifth Generation Computer Systems*, ICOT, pp. 693-702, 1988.

14. DUIJVESTIJN A. J. W., Simple Perfect Squared Square of Lowest Order, *Journal of Combinatorial Theory*, Series B 25, pp. 240-243, 1978.

15. GARDNER M., Wheels, life and other mathematical amusements, W. H. Freeman and Compagny, 1983.

16. GENESERETH M. R. and M. L. GINSBERG, Logic Programming, *Communications of the ACM*, Volume 28, Number 9, 933-941, September 1985.

17. IMBERT J-L., About Redundant Inequalities Generated by Fourier's Algorithm, AIMSA'90, *4th International Conference on Artificial Intelligence : Methodology, Systems, Applications*, Albena-Varna, Bulgaria, September 1990, Proceedings to be published by North-Holland,

18. JAFFAR J. and J-L. LASSEZ, Constraint Logic Programming, *14th ACM Symposium on the Principle of Programming languages*, pp. 111-119, 1987.

19. JAFFAR J. and S. MICHAYLOV, Methodology and Implementation of a Constraint Logic Programming System, *Proceedings of the Fourth International Conference on Logic Programming*, Melbourne, MIT Press, pp. 196-218, 1987.

20. KOWALSKI R. and D. KUEHNER, Resolution with Selection Function, *Artificial Intelligence*, Vol. 3, No. 3, pp. 227-260, 1970.

21. OXUSOFF L. and A. RAUZY., *Evaluation sémantique en calcul propositionnel*, Thèses de Doctorat, GIA, Faculté des Sciences de Luminy, Université Aix-Marseille II, January 1989.

22. ROBINSON A., A machine-oriented logic based on the resolution principle, *Journal of the ACM*, 12, December 1965.

23. TOURAIVANE, *La récupération de mémoire dans les machines non déterministes*, Thèse de Doctorat, Faculté des Sciences de Luminy, Université Aix-Marseille II, November 1988.

24. SIEGEL P., *Représentation et utilisation de la connaissance en calcul propositionnel*, Thèse de Doctorat d'Etat, GIA, Faculté des Sciences de Luminy, Université Aix-Marseille II, July 1987.

25. SPRAGUE R., Über die Zerlegung von Rechtecken in lauter verschiedene Quadrate, *Journal für die reine und angewandte Mathematik*, 182, 1940.

On Open Defaults

Vladimir Lifschitz

Stanford University, Stanford, California 94305, USA
and
University of Texas, Austin, Texas 78712, USA

Abstract

In Reiter's default logic, the parameters of a default are treated as metavariables for ground terms. We propose an alternative definition of an extension for a default theory, which handles parameters as genuine object variables. The new form of default logic may be preferable when the domain closure assumption is not postulated. It stands in a particularly simple relation to circumscription. Like circumscription, it can be viewed as a syntactic transformation of formulas of higher order logic.

1 Introduction

Default logic [Reiter, 1980] is one of the most expressive and most widely used nonmonotonic formalisms. In one respect, however, the main definition of default logic, that of an *extension*, is not entirely satisfactory.

Recall that a default

$$\alpha : \beta_1, \ldots, \beta_m / \gamma \qquad (1)$$

is *open* if it contains free variables, and *closed* otherwise. The concept of an extension is defined in two steps: It is first introduced, by means of a fixpoint construction, for default theories without open defaults, and then generalized to arbitrary default theories. Since interesting cases usually involve open defaults, the second step is crucial. Its main idea is that a default with free variables has the same meaning as the set of all its ground instances.[1] In other words, free variables in a default are viewed as metavariables for ground terms.

In many cases, this treatment of free variables makes the effect of a default surprisingly weak. Consider the default theory with one axiom $P(a)$ and one default,

$$: \neg P(x) / \neg P(x). \qquad (2)$$

Intuitively, this default expresses that $P(x)$ is assumed to be false whenever possible. We can expect that it will allow us to prove

$$\forall x (P(x) \equiv x = a). \qquad (3)$$

---

[1]The actual reduction of the general case to the case of closed defaults is more complex, because it involves the Skolemization of all axioms and of the consequents of all defaults—a detail which is irrelevant for this discussion.

But all that this default gives is the literals $\neg P(t)$ for the ground terms t different from a. Notice that the behavior of circumscription[2] is quite different. The circumscription of P in $P(a)$, which expresses the same idea of making $P(x)$ false whenever possible, is equivalent to (3).

In applications to formalizing commonsense reasoning, this weakness of open defaults is sometimes undesirable. Consider the following example.[3] Suppose that, for any two blocks x and y, the default is that x is not on y. If there is no evidence that any blocks are on the block B_1 then, for each individual block B_i, we will be able to conclude that it is not on B_1. But we may be unable to justify the conclusion that B_1 is clear, in the sense that there are no blocks on B_1. Indeed, the set of conclusions

$$\neg on(B_1, B_1), \ldots, \neg on(B_n, B_1)$$

is weaker than the universally quantified formula

$$\forall x \neg on(x, B_1),$$

unless we accept the "domain closure assumption"

$$\forall x (x = B_1 \vee \ldots \vee x = B_n),$$

expressing that every block is represented by one of the constants B_i. The domain closure assumption is sometimes unacceptable: We may be unable or unwilling to design the language in such a way that each object in the domain of reasoning be represented by a ground term.

In this paper we propose a modification of default logic in which free variables in defaults are treated as genuine object variables, rather than metavariables for ground terms. The new form of default logic is better suited for formalizing default reasoning in the absence of the domain closure assumption. Another reason why this modification of Reiter's system can be of interest is that it stands in a particularly simple relation to circumscription and consequently sheds some light on the important and difficult problem of connecting various approaches to default reasoning.

As our starting point, we take a characterization of extensions for the case of closed defaults based on [Guerreiro and Casanova, 1990]. This characterization uses a fixpoint construction which is similar to Reiter's, except that the fixpoints in question are *classes of models*, rather than sets of sentences. It is not particularly surprising that extensions can be defined in such a manner, because there is a natural correspondence between classes of models and sets of sentences: For any class V of models, we can consider its *theory*, that is, the set of sentences that are true in all models from V. But for our purposes the transition from formulas to models is essential. In the realm of formulas, the only way to refer to objects in the domain of reasoning is through their syntactic representations, that is, ground terms. In the realm of models, we can talk about elements of the universe directly.

In Section 2 we review, for motivation and further reference, the fixpoint constructions from [Reiter, 1980] and [Guerreiro and Casanova, 1990]. The new definition of an

---

[2]The definition of circumscription can be found in [McCarthy, 1986] or [Lifschitz, 1985].

[3][McCarthy, 1980], Section 8, Remark 2; [Poole, 1987b], Example 3. See also the discussion of "default reasoning in an open domain" (the note to Example A5) in [Lifschitz, 1989a].

extension is given in Section 3. Then we discuss its relation to traditional default logic (Section 4) and to circumscription (Section 5). In Section 6 we outline an extension of the formalism in which some object, function or predicate constants are treated as "fixed." In Section 7 we show that the new form of default logic, like circumscription, can be viewed as a syntactic transformation of formulas of higher order logic. Related work is surveyed in Section 8. Most proofs are deferred to the appendix.

2   Extensions According to Reiter and Guerreiro–Casanova

According to [Reiter, 1980], a *default theory* (in a given first order language) is a pair (D, W), where D is a set of defaults of the form (1), and W is a set of sentences. A default theory (D, W) is *closed* if all defaults from D are closed. Let (D, W) be a closed default theory. For any set of sentences S, consider the smallest set of sentences S' which includes W, is closed under classical logic, and satisfies the condition:

(∗) For any default (1) from D, if $\alpha \in S'$ and $\neg\beta_1, \ldots, \neg\beta_m \notin S$ then $\gamma \in S'$.

This set S' is denoted by $\Gamma(S)$. S is said to be an *extension* for (D, W) if it is a fixpoint of Γ, that is, if $\Gamma(S) = S$.

The Guerreiro–Casanova approach to closed default theories can be described as follows. For any class V of structures for the language of (D, W), let $Th(V)$ stand for the theory of V—the set of sentences which are true in all structures from V. Let V' be the largest class of models of W which satisfies the condition:

(∗∗) For any default (1) from D, if $\alpha \in Th(V')$ and $\neg\beta_1, \ldots, \neg\beta_m \notin Th(V)$ then $\gamma \in Th(V')$.

This largest V' always exists:

Proposition 1. *The union of all classes V' of models of W which satisfy* (∗∗) *satisfies* (∗∗) *also.*

This class V' is denoted by $\Sigma(V)$. As essentially established in [Guerreiro and Casanova, 1990], extensions can be characterized in terms of the fixpoints of Σ:

Proposition 2. *A set of sentences is an extension for (D, W) if and only if it has the form $Th(V)$ for some fixpoint V of Σ.*

Intuitively, if we think of the extension $Th(V)$ as a set of "beliefs," then V is the class of "worlds" that are possible according to these beliefs.

Our objective is to generalize the definition of Σ to open defaults. Then the characterization of extensions given by Proposition 2 will serve as the basis for a new definition of an extension.

Consider an open default

$$\alpha(x) : \beta_1(x), \ldots, \beta_m(x)/\gamma(x), \tag{4}$$

where x is a list of variables. Instead of replacing the variables from x by ground terms, as in Reiter's logic, we want to replace them by arbitrary elements of the universe, or, more precisely, by symbols that serve as "names" of arbitrary elements of the universe. The condition (∗∗) will turn into something like this: For any default (4) from D and any tuple ξ of "names," if $\alpha(\xi) \in Th(V')$ and $\neg\beta_1(\xi), \ldots, \neg\beta_m(\xi) \notin Th(V)$ then $\gamma(\xi) \in Th(V')$.

There is a problem, however, with this idea: Different structures from V have, generally, different universes, and "names" appropriate for one structure from V will generally make no sense for another. In order to generalize the condition (∗∗) to open defaults, we need to modify it so that all structures from V have the same universe.

3 Default Logic with a Fixed Universe

Let (D, W) be a default theory, not necessarily closed, and let U be a nonempty set. By a *world* we understand any model of W with the universe U. Extend the language of (D, W) by object constants representing all elements of U; these constants will be called *names*. For any set of worlds V, $Th^*(V)$ is the set of sentences in the extended language which are true in all worlds from V. (Thus $Th(V)$ is the set of the sentences from $Th^*(V)$ that do not contain names.)

For any set of worlds V, consider the largest set of worlds V' which satisfies the condition:

(∗∗∗) For any default (4) from D and any tuple of names ξ, if $\alpha(\xi) \in Th^*(V')$ and $\neg\beta_1(\xi), \ldots, \neg\beta_m(\xi) \notin Th^*(V)$ then $\gamma(\xi) \in Th^*(V')$.

This largest V' always exists:

Proposition 3. *The set of sets V' satisfying* (∗∗∗) *is closed under union.*

This set V' will be denoted by $\Delta(V)$. The operator Δ is the "fixed universe" counterpart of Σ. Notice that Δ depends not only on the default theory (W, D), but also on the universe U.

A *U-extension* for (D, W) is any set of sentences of the form $Th(V)$, where V is a fixpoint of Δ. Notice that U-extensions, just like extensions in Reiter's logic, consist of sentences in the language of (D, W); they do not contain names.

It is clear that the U-extensions for a given default theory (D, W) are completely determined by the cardinality of U. For any positive integer n, let $Card_n$ be a standard sentence expressing that there are exactly n objects; for instance, we can take $Card_1$ to be $\forall xy(x = y)$. By $Card_U$ we denote $Card_n$ if the cardinality of U is n, and $\{\neg Card_1, \neg Card_2, \ldots\}$ if U is infinite. Any U-extension contains W and $Card_U$, and is closed under classical logic.

As an example, consider the operator Δ for the default theory discussed in the introduction:

$$W = \{P(a)\},\ D = \{: \neg P(x)/\neg P(x)\}. \tag{5}$$

We will see in Section 5 that, if M is a model of (3) with the universe U, then $\{M\}$ is a fixpoint of Δ, and, conversely, every fixpoint of Δ has this form. The U-extension $Th(V)$ corresponding to a fixpoint V of Δ is the deductive closure of (3) and $Card_U$. Thus, for every U, there is exactly one U-extension.

In Reiter's logic, we can say that a sentence A is a *consequence* of (D, W) if A is in the intersection of all extensions for (D, W). For instance, the consequences of the default theory (5) are the logical consequences of $P(a)$. Let us say that A is a *fixed universe consequence (F-consequence)* of (D, W) if A is in the intersection of all U-extensions for (D, W) over all nonempty sets U. In other words, an F-consequence is a sentence which is true in every world which belongs to some fixpoint of Δ. It is clear that the set of F-consequences of (D, W), like the set of its consequences, contains W and is closed under classical logic. In case of (5), the F-consequences are the logical consequences of (3), which is exactly what we wanted to achieve.

If D is empty, then the condition $(***)$ is trivially true, so that, for every V, $\Delta(V)$ is the set of all worlds, and this set is the only fixpoint of Δ. Consequently, the only U-extension of $(\emptyset, W)$ consists of the sentences that are true in all models of W with the universe U. It follows that the F-consequences of this theory are the sentences logically entailed by W.

4   Relation to Reiter's Logic

Often, as in the example (5), a default theory has more F-consequences than consequences. But sometimes this is the other way around. Consider the following example:

$$W = \{P(a)\}, \; D = \{: \neg P(b)/\neg P(b)\}. \tag{6}$$

The extension of this theory in Reiter's logic includes $\neg P(b)$. If, on the other hand, the cardinality of U is 1, then the U-extension of this theory includes $P(b)$ (which is a logical consequence of W and $Card_1$), and does not include $\neg P(b)$. Consequently, $\neg P(b)$ is not an F-consequence of (6). The sentence $a \neq b$ is another consequence of (6) which is not an F-consequence.

This example shows also that the two versions of default logic are not equivalent even for closed default theories. For closed defaults, $(***)$ turns into $(**)$, except that both V and V' are assumed to consist of models with a fixed universe U; this distinction is responsible for the difference between consequences and F-consequences.

If, however, the language is propositional, then the choice of U becomes inessential, because models of a propositional theory are simply mappings of propositional symbols into truth values. Consequently, for propositional default theories, Δ coincides with Σ. It follows then by Proposition 2, that, in the propositional case, U-extensions are identical to extensions, and F-consequences are identical to consequences.

Here are some other cases when the two forms of default logic are equivalent:

Proposition 4. *Let (D, W) be a closed default theory. If all models of W have the same finite cardinality, then the F-consequences of (D, W) are identical to its consequences.*

Proposition 5. *Let (D, W) be a closed default theory with at most countably many object, function and predicate constants. If all models of W are infinite, then the F-consequences of (D, W) are identical to its consequences.*

Proposition 4 is applicable when W includes both the domain closure assumption and the unique names assumption. Proposition 5 is applicable when the universe of discourse includes some infinite domain, for instance, natural numbers.

As an illustration of Proposition 5, we can consider a modification of (6) in which axioms expressing the existence of infinitely many objects are added to W. Such a theory will have both $\neg P(b)$ and $a \neq b$ among its F-consequences. Even simpler, this effect can be achieved by assuming two distinct objects:

$$W = \{P(a), \exists xy (x \neq y)\}, \ D = \{: \neg P(b)/\neg P(b)\}.$$

If U is a singleton, then the set of worlds for this theory is empty, and $\emptyset$ is the only fixpoint of Δ. Consequently, the only U-extension for a singleton U is the set of all sentences, and the argument made at the beginning of this section regarding (6) does not go through.

5   Normal Defaults Without Prerequisites

In this section we assume that D is

$$\{: \alpha(x)/\alpha(x)\}, \tag{7}$$

so that it consists of a single default, which is a "normal default without a prerequisite." For example, each of the theories (5), (6) satisfies this condition. Intuitively, the effect of the default (7) is to "maximize" α. Proposition 6 below gives a characterization of the U-extensions for such theories which makes this claim precise.

The following notation will be used: For any world M, α^M stands for the set of tuples of names ξ such that $\alpha(\xi)$ is true in M.

Let us say that a world M is α-*maximal* if there is no world M' such that α^M is a proper subset of $\alpha^{M'}$. In other words, an α-maximal world is a model of W with the universe U such that no other model of W with the same universe has more x's satisfying $\alpha(x)$. In particular, if $\alpha(x)$ is $\neg P(x)$, then an α-maximal world is a model in which the extent of P is *minimal* in the sense corresponding to the circumscription of P with all object, function and predicate constants allowed to vary.

In the following proposition we assume that the set of worlds is nonempty, that is, W has at least one model with the universe U.

Proposition 6. *For any default theory with the set of defaults (7), every fixpoint of the corresponding operator Δ is an equivalence class of the set of α-maximal worlds relative to the relation $\alpha^M = \alpha^{M'}$. Conversely, each of these equivalence classes is a fixpoint of Δ.*

This theorem shows that the fixpoints of Δ correspond to the extents of α in α-maximal worlds. The claims made above about the default theories (5) and (6) can be easily justified using Proposition 6. For (5), the α-maximal worlds are the models of (3), and each equivalence class consists of a single model.

Proposition 6 implies that the F-consequences of a theory with the set of defaults (7) can be characterized as the sentences that are true in all α-maximal worlds for all U. In particular:

Corollary. *A sentence $B(P)$ is an F-consequence of the default theory*

$$(\{A(P)\}, \{: \neg P(x)/\neg P(x)\})$$

if and only if B(P) is logically entailed by the circumscription of P in A(P) with all object, function and predicate constants allowed to vary.

In the corresponding result for Reiter's default logic ([Etherington, 1987a], Theorem 2), W is required to include the domain closure assumption and a form of the unique names assumption. The new approach to open defaults makes these conditions redundant.

6 Default Logic with Fixed Constants

As we have seen, circumscription with all constants varied is a special case of default logic with a fixed universe. In this section, a generalization of this form of default logic is defined, which is related to circumscription with some object, function and predicate constants fixed. This generalization provides a new perspective on the relationship between circumscription and default logic. We do not propose it as a serious candidate for AI use.

A *default theory with fixed constants* is a triple (D, W, C), where D and W are as in the standard definition of a default theory, and C is a subset of object, function and predicate constants. The symbols from C are the *fixed* constants of the theory; the remaining constants are *varied*. A default theory (D, W) corresponds to the case when C is empty. By L_C we denote the first order language whose object, function and predicate constants are the members of C; thus L_C is a sublanguage of the language of the theory.

The definitions of the operator Δ and of a U-extension for this generalization of default theories are the same as in Section 3, except that now we take U to be a structure for the language L_C (rather than merely a universe), and define a *world* to be any model of W obtained from U by assigning interpretations to the varied constants. It is clear that the U-extensions for a default theory with fixed constants remain the same if U is replaced by an isomorphic structure.

An *F-consequence* of a default theory with fixed constants is a sentence that belongs to all its U-extensions for all structures U.

Consider, for instance, the default theory with

$$W = \{\forall x(Q(x) \supset P(x))\}, \; D = \{: \neg P(x)/\neg P(x)\}, \; C = \{Q\}. \tag{8}$$

Let U be a structure for the language whose only nonlogical constant is Q (that is, U is a universe along with its subset representing Q). A world is defined by an interpretation of P that makes the sentence $\forall x(Q(x) \supset P(x))$ true, that is, by a subset of the universe that contains the set representing Q. The only fixpoint of Δ consists of one world, in which the extent of P is the same as the extent of Q. The F-consequences of (8) are the sentences logically entailed by

$$\forall x(P(x) \equiv Q(x)). \tag{9}$$

The definition of an α-maximal world (Section 5), Proposition 6 and its proof apply to default theories with fixed constants, without any changes whatsoever. The counterpart of Corollary to Proposition 6 can be stated as follows:

Proposition 7. *A sentence B(P) is an F-consequence of the theory*

$$(\{A(P)\}, \{: \neg P(x)/\neg P(x)\}, C),$$

where C does not include P, if and only if $B(P)$ is logically entailed by the circumscription of P in $A(P)$ with the fixed constants C.

We see that default theories with fixed constants subsume a rather general form of circumscription. The fact that the F-consequences of (8) are the sentences logically entailed by (9) can serve as an illustration of this theorem, because (9) is the result of circumscribing P in $\forall x(Q(x) \supset P(x))$ with Q fixed.

7 Default Logic as a Syntactic Transformation

Our next goal is to show that the definition of a U-extension can be expressed by a higher order logical formula, so that the modification of default logic proposed in this paper, like circumscription, can be viewed as a syntactic transformation of sentences. Since the definition of a U-extension involves not only worlds, but also *sets* of worlds, we will need not only second order, but also *third order* variables.

Let (D, W, C) be a default theory with fixed constants, in a language which has finitely many object, function and predicate constants, and with both D and W finite. Let Z be the list of all varied constants (that is, the constants that do not belong to C). We will explicitly show the occurences of the varied constants in formulas, so that an arbitrary sentence will be written as $F(Z)$, and a formula with the list of parameters x will be written as $F(Z, x)$. The set W will be identified with the conjunction of its elements and written as $W(Z)$.

Take a list of variables z of the same length as Z, such that if the i-th member of Z is an object constant then the i-th member of z is an object variable, and if the i-th member of Z is a function (predicate) constant then the i-th member of z is a function (predicate) variable of the same arity. Given a structure U for the language L_C, the structures obtained from U by assigning interpretations to the varied constants can be identified with combinations of values of the variables z in U. In particular, the models of W that are obtained in this way ("worlds") correspond to the values of z for which $W(z)$ is true.

Let v be a variable such that $v(z)$ is a well-formed formula (so that v is third order if Z contains at least one function or predicate constant). Values of v can be identified with sets of structures obtained from U by assigning interpretations to the varied constants. Then the values of v satisfying the condition

$$\forall z[v(z) \supset W(z)] \tag{10}$$

represent sets of worlds. More generally, for any formula $F(Z, x)$, the formula

$$\forall z[v(z) \supset F(z, x)]$$

expresses that $F(Z, x)$ is true in each structure from v. We will denote this formula by $F^\forall(v, x)$, so that (10) will be written as $W^\forall(v)$. Similarly, the formula

$$\exists z[v(z) \wedge F(z, x)],$$

expressing that $F(Z, x)$ is true in at least one structure from v, will be denoted by $F^\exists(v, x)$.

Using this notation, it is easy to encode the relation between sets of worlds V, V' expressed by (***) (Section 3). If d is a default

$$\alpha(Z,x) : \beta_1(Z,x),\ldots,\beta_m(Z,x)/\gamma(Z,x)$$

from D, then by $d(v,v')$ we denote the formula

$$\forall x[\alpha^\forall(v',x) \wedge \beta_1^\exists(v,x),\ldots,\beta_m^\exists(v,x) \supset \gamma^\forall(v',x)].$$

Then the condition (***) is expressed by the conjunction

$$\bigwedge_{d \in D} d(v,v').$$

Since $\Delta(V)$ is the union of all sets V' of worlds satisfying (***), it is the represented by the abbreviation

$$\Delta(v) = \lambda z \exists v'[W^\forall(v') \wedge \bigwedge_{d \in D} d(v,v') \wedge v'(z)].$$

Now we can define the "default logic operator" DL. By $DL(D,W,C)$ we denote the sentence

$$\exists v[\Delta(v) = v \wedge v(Z)].$$

A structure obtained from U by assigning values to the varied constants is a model of this sentence if and only if it belongs to some fixpoint of the corresponding operator Δ. Consequently, the class of models of $DL(D,W,C)$ is the union of all fixpoints of the operators Δ corresponding to all structures U. We conclude:

Proposition 8. *A sentence B is an F-consequence of (D,W,C) if and only if B is logically entailed by $DL(D,W,C)$.*

8 Related Work

The correspondence between extensions and classes of models on which our main definition is based was first used by Etherington [1987b], although his construction is less transparent than that of [Guerreiro and Casanova, 1990].

The possibility of introducing "nonground instances of defaults" is discussed by Poole [1987b] for his formulation of default logic, equivalent to the normal subset of Reiter's system. In [Poole, 1987a], the distinction between varied and fixed predicates is added to that formalism.

Przymusinski [1989] defends the use of non-Herbrand models in logic programming. Since logic programs can be viewed as a special case of default theories [Bidoit and Froidevaux, 1988], this issue is related to the problem of open defaults.

The counterpart of an open default theory in autoepistemic logic is an autoepistemic theory with "quantifying-in." One way of defining a semantics for such theories is proposed by Levesque [1990]. Konolige [1989] discusses another approach; he also studies the problem of reducing circumscription to autoepistemic logic without assuming domain closure. Introspective circumscription [Lifschitz, 1989b] is a system analogous to autoepistemic logic, in which unrestricted quantification is allowed. Like the formalism of this paper, it subsumes some forms of "minimizing" circumscription.

Levesque's formulation of autoepistemic logic is based on a mapping into a monotonic system, like the characterization of our default logic given in Section 7. One difference is that Levesque uses a simple mapping into a rather involved modal logic, and we use a rather complicated transformation whose target language is classical.

9 Conclusion

Recent research shows that the main ideas of different nonmonotonic formalisms are more compatible with each other than one might think. Perhaps we will not have to select any one of the classical approaches ([McCarthy, 1980], [McDermott and Doyle, 1980], [Reiter, 1980]) as a basis for the nonmonotonic formalism of the future, and reject the others; we may be able to combine the advantages of different models of nonmonotonic reasoning in the same system.

Konolige [1988] noticed, for instance, that modal nonmonotonic languages, such as autoepistemic logic, are close in their expressiveness to the language of default logic. Unfortunately, the problem of finding precise equivalence results for Konolige's translation turned out to be quite difficult. Several "groundedness" conditions have been proposed in order to filter out the autoepistemic extensions that have no counterparts in default logic. The first attempt [Konolige, 1988] was unsuccessful, and this has led to the invention of "supergrounded" and "robust" extensions (see ([Marek and Truszczyński, 1989], Sections 2.3 and 3). Further work in this direction is described in [Marek and Truszczyński, 1990].

The available reductions of defaults to epistemic formulas are not completely satisfactory. What makes the situation even more complicated is the fact that autoepistemic logic is merely one point in the whole spectrum of nonmonotonic modal systems introduced in [McDermott, 1982], and possibly not the best for AI applications [Shvarts, 1990]. On the other hand, the availability of an epistemic modal operator is an attractive feature of a knowledge representation language, in connection with the problem of representing integrity constraints [Reiter, 1988]. Hopefully, future research will lead to the invention of an elegant modal system, such that default logic will be linked to it by means of Konolige's translation or a similar mechanism.

The results of this paper suggest that such a system may very well be a superset of some forms of circumscription. It is also possible that this system, like circumscription, will be defined by a syntactic transformation with a clear model-theoretic meaning—and this is what makes the definition of circumscription so attractive in the first place.

Since some logic programming languages with negation as failure can be easily embedded into default logic ([Bidoit and Froidevaux, 1988], [Gelfond and Lifschitz, 1990], [Kowalski and Sadri, 1990]), we can expect that they, too, will become subsets of the nonmonotonic system of the future. These subsets will be important in view of their good computational properties. It may be possible to automate reasoning in more complex nonmonotonic theories by compiling them into logic programs [Gelfond and Lifschitz, 1989] or by constructing tractable "approximations" to them.

Acknowledgments

I would like to thank David Etherington, Michael Gelfond, Hector Levesque, John McCarthy and Ramiro Guerreiro for comments on earlier versions of this paper. This research was supported in part by NSF grant IRI-8904611 and by DARPA under Contract N00039-84-C-0211.

References

[Bidoit and Froidevaux, 1988] Nicole Bidoit and Christine Froidevaux. Negation by default and nonstratifiable logic programs. Technical Report 437, Université Paris XI, September 1988.

[Etherington, 1987a] David Etherington. Relating default logic and circumscription. In *Proc. IJCAI-87*, pages 489–494, 1987.

[Etherington, 1987b] David Etherington. A semantics for default logic. In *Proc. IJCAI-87*, pages 495–498, 1987.

[Gelfond and Lifschitz, 1989] Michael Gelfond and Vladimir Lifschitz. Compiling circumscriptive theories into logic programs. In Michael Reinfrank, Johan de Kleer, Matthew Ginsberg, and Erik Sandewall, editors, *Non-Monotonic Reasoning: 2nd International Workshop (Lecture Notes in Artificial Intelligence 346)*, pages 74–99. Springer-Verlag, 1989.

[Gelfond and Lifschitz, 1990] Michael Gelfond and Vladimir Lifschitz. Logic programs with classical negation. In David Warren and Peter Szeredi, editors, *Logic Programming: Proc. of the Seventh Int'l Conf.*, pages 579–597, 1990.

[Guerreiro and Casanova, 1990] Ramiro Guerreiro and Marco Casanova. An alternative semantics for default logic. Preprint, The Third International Workshop on Nonmonotonic Reasoning, South Lake Tahoe, 1990.

[Konolige, 1988] Kurt Konolige. On the relation between default and autoepistemic logic. *Artificial Intelligence*, 35:343–382, 1988.

[Konolige, 1989] Kurt Konolige. On the relation between autoepistemic logic and circumscription. In *Proc. of IJCAI-89*, pages 1213–1218, 1989.

[Kowalski and Sadri, 1990] Robert Kowalski and Fariba Sadri. Logic programs with exceptions. In David Warren and Peter Szeredi, editors, *Logic Programming: Proc. of the Seventh Int'l Conf.*, pages 598–613, 1990.

[Levesque, 1990] Hector Levesque. All I know: a study in autoepistemic logic. *Artificial Intelligence*, 42(2,3):263–310, 1990.

[Lifschitz, 1985] Vladimir Lifschitz. Computing circumscription. In *Proc. of IJCAI-85*, pages 121–127, 1985.

[Lifschitz, 1989a] Vladimir Lifschitz. Benchmark problems for formal non-monotonic reasoning, version 2.00. In Michael Reinfrank, Johan de Kleer, Matthew Ginsberg, and Erik Sandewall, editors, *Non-Monotonic Reasoning: 2nd International Workshop (Lecture Notes in Artificial Intelligence 346)*, pages 202–219. Springer-Verlag, 1989.

[Lifschitz, 1989b] Vladimir Lifschitz. Between circumscription and autoepistemic logic. In Ronald Brachman, Hector Levesque, and Raymond Reiter, editors, *Proc. of the First Int'l Conf. on Principles of Knowledge Representation and Reasoning*, pages 235–244, 1989.

[Marek and Truszczyński, 1989] Wiktor Marek and Miroslaw Truszczyński. Relating autoepistemic and default logic. In Ronald Brachman, Hector Levesque, and Raymond Reiter, editors, *Proc. of the First Int'l Conf. on Principles of Knowledge Representation and Reasoning*, pages 276–288, 1989.

[Marek and Truszczyński, 1990] Wiktor Marek and Miroslaw Truszczyński. Modal logic for default reasoning. To appear, 1990.

[McCarthy, 1980] John McCarthy. Circumscription—a form of non-monotonic reasoning. *Artificial Intelligence*, 13(1, 2):27–39,171–172, April 1980.

[McCarthy, 1986] John McCarthy. Applications of circumscription to formalizing common sense knowledge. *Artificial Intelligence*, 26(3):89–116, 1986.

[McDermott and Doyle, 1980] Drew McDermott and Jon Doyle. Nonmonotonic logic I. *Artificial Intelligence*, 13(1,2):41–72, 1980.

[McDermott, 1982] Drew McDermott. Nonmonotonic logic II: Nonmonotonic modal theories. *Journal of the ACM*, 29(1):33–57, 1982.

[Poole, 1987a] David Poole. Fixed predicates in default reasoning. Manuscript, 1987.

[Poole, 1987b] David Poole. Variables in hypotheses. In *Proc. IJCAI-87*, pages 905–908, 1987.

[Przymusinski, 1989] Teodor Przymusinski. On the declarative and procedural semantics of logic programs. *Journal of Automated Reasoning*, 5:167–205, 1989.

[Reiter, 1980] Raymond Reiter. A logic for default reasoning. *Artificial Intelligence*, 13(1,2):81–132, 1980.

[Reiter, 1988] Raymond Reiter. On integrity constraints. In Moshe Vardi, editor, *Theoretical Aspects of Reasoning about Knowledge: Proc. of the Second Conf.*, pages 97–111, 1988.

[Shvarts, 1990] Grigori Shvarts. Autoepistemic modal logics. In Rohit Parikh, editor, *Theoretical Aspects of Reasoning about Knowledge: Proc. of the Third Conf.*, pages 97–110, 1990.

Appendix. Proofs of Theorems

Proposition 1. *The union of all classes V' of models of W which satisfy $(**)$ satisfies $(**)$ also.*

Proof. Let V_0' be the union of all classes V' of models which satisfy $(**)$, and let (1) be a default from D such that $\alpha \in Th(V_0')$ and $\neg\beta_1, \ldots, \neg\beta_m \notin Th(V)$. Take any model $M \in V_0'$. By the choice of V_0', there exists a class V' of models which satisfies the conditions $(**)$ and $M \in V' \subset V_0'$. Since $\alpha \in Th(V_0') \subset Th(V')$, we can conclude that $\gamma \in Th(V')$. Consequently γ is true in M. Thus γ is true in every model from V_0', that is, $\gamma \in Th(V_0')$.

The proof of Proposition 2 is based on two lemmas.

Lemma 1. *For any class of structures V, $\Gamma(Th(V)) = Th(\Sigma(V))$.*

Proof. According to the definition of Γ, $\Gamma(Th(V))$ is the smallest class of sentences S' such that (i) S' is closed under classical logic, (ii) $W \subset S'$, and (iii) for any default (1) from D, if $\alpha \in S'$ and $\neg\beta_1, \ldots, \neg\beta_m \notin Th(V)$ then $\gamma \in S'$. Notice that a set of sentences S' is closed under classical logic if and only if it can be represented in the form $Th(V')$ for some class of structures V'. (Proof: Take V' to be the class of all models of S'.) Furthermore, $W \subset Th(V')$ means that every member of V' is a model of W. Consequently, $\Gamma(Th(V))$ can be characterized as the smallest class of sentences of the form $Th(V')$, where V' is a class of models of W, such that $(**)$ holds. Since the operator Th is monotone decreasing, this is the same as $Th(V')$ for the largest class V' of models of W satisfying $(**)$, that is, the same as $Th(\Sigma(V))$.

Lemma 2. *The class of models of $Th(\Sigma(V))$ coincides with $\Sigma(V)$.*

Proof. If V_1' is any class of models of W which satisfies $(**)$, and V_2' is any class of structures such that $Th(V_2') = Th(V_1')$, then V_2' is a class of models of W, and it satisfies $(**)$ also. By applying this to $\Sigma(V)$ as V_1' and to the class of models of $Th(\Sigma(V))$ as V_2', we conclude that the class of models of $Th(\Sigma(V))$ is a class of models of W and satisfies $(**)$. Since it contains $\Sigma(V)$, which is the largest such class, the two classes coincide.

Proposition 2. *A set of sentences is an extension for (D, W) if and only if it has the form $Th(V)$ for some fixpoint V of Σ.*

Proof. If V is a fixpoint of Σ, then, by Lemma 1,

$$\Gamma(Th(V)) = Th(\Sigma(V)) = Th(V),$$

so that $Th(V)$ is an extension. To show that any extension can be represented in this form, consider any extension S, and let V be the class of its models. Since S is closed under first-order logic, $S = Th(V)$, and it remains to check that V is a fixpoint of Σ. Since S is a fixpoint of Γ, $Th(V) = \Gamma(Th(V))$. From the last two equalities and Lemma 1, $S = Th(\Sigma(V))$. By Lemma 2, it follows that $\Sigma(V)$ is class of models of S, that is, $\Sigma(V) = V$.

Proposition 3. *The set of sets V' satisfying $(***)$ is closed under union.*

Proof. Let V_0' be the union of some of the sets V' which satisfy $(***)$, let (4) be a default from D, and let ξ be a tuple of names such that $\alpha(\xi) \in Th^*(V_0')$ and $\neg\beta_1(\xi), \ldots, \neg\beta_m(\xi) \notin$

$Th^*(V)$. Take any world M from V_0'. By the choice of V_0', there exists a set V' of worlds which satisfies the conditions $(***)$ and $M \in V' \subset V_0'$. Since $\alpha(\xi) \in Th^*(V_0') \subset Th^*(V')$, we can conclude that $\gamma(\xi) \in Th^*(V')$. Consequently $\gamma(\xi)$ is true in M. Thus $\gamma(\xi)$ is true in every world from V_0', that is, $\gamma(\xi) \in Th^*(V_0')$.

The proofs of Propositions 4 and 5 use the following terminology. Given a set of sentences W and a nonempty set U, we say that U is W-complete if for every model M of W there exists a structure with the universe U that is elementarily equivalent to M. For instance, if all models of W have the same cardinality, then any set of this cardinality is W-complete. The Löwenheim—Skolem theorem shows that if W has at most countably many constants and no finite models, then any infinite set is W-complete.

For any structure M, let $\overline{M}$ be the class of structures elementarily equivalent to M. If V is a class of structures, then $\overline{V}$ stands for $\bigcup_{M \in V} \overline{M}$. Obviously, $V \subset \overline{V}$ and $\overline{\overline{V}} = \overline{V}$.

By V_0 we will denote the set of all worlds (that is, of all models of W with the universe U).

Lemma 3. *If U is W-complete, then, for any class V of models of W,*

$$\overline{\overline{V} \cap V_0} = \overline{V}.$$

Proof. The inclusion left to right is obvious. Take any model $M \in \overline{V}$, and let M' be a world elementarily equivalent to M. Then

$$M' \in \overline{M} \cap V_0 \subset \overline{V} \cap V_0.$$

Consequently,

$$M \in \overline{M'} \subset \overline{\overline{V} \cap V_0}.$$

For any class V of models of W,

$$\overline{\Sigma(V)} = \Sigma(V), \tag{11}$$

because $\Sigma(V)$ is the largest class satisfying $(**)$, and $\overline{\Sigma(V)}$ is its superclass with the same theory. Observe also that, if U is W-complete, then

$$Th(V) = Th(\overline{V}) = Th(\overline{V} \cap V_0). \tag{12}$$

Since $\Sigma(V)$ is invariant with respect to replacing V by another class of structures with the same theory, it follows that

$$\Sigma(V) = \Sigma(\overline{V}) = \Sigma(\overline{V} \cap V_0). \tag{13}$$

Lemma 4. *For any closed default theory (D, W), any W-complete U, and any set of worlds V,*

$$\Delta(V) = \Sigma(V) \cap V_0.$$

Proof. It is clear that $\Delta(V)$ is contained both in $\Sigma(V)$ and in V_0. In order to prove that $\Sigma(V) \cap V_0$ is a subset of $\Delta(V)$, we only need to check that it satisfies $(**)$ as V'. This follows from the fact that, by (11) and (12), $\Sigma(V) \cap V_0$ has the same theory as $\Sigma(V)$.

Lemma 5. *For any closed default theory* (D, W), *any* W-*complete* U, *and any set of worlds* V,

$$\Sigma(V) = \overline{\Delta(V)}.$$

Proof. Obviously, $\Delta(V) \subset \Sigma(V)$. Then, by (11),

$$\overline{\Delta(V)} \subset \overline{\Sigma(V)} = \Sigma(V).$$

To prove that $\Sigma(V)$ is a subclass of $\overline{\Delta(V)}$, we need to check that, for any V' satisfying (∗∗), $V' \subset \overline{\Delta(V)}$. Take any V' satisfying (∗∗). By (12), $Th(\overline{V'} \cap V_0) = Th(V')$. Consequently, $\overline{V'} \cap V_0$ satisfies (∗∗) also, so that $\overline{V'} \cap V_0 \subset \Delta(V)$. Then, by Lemma 3, $V' \subset \overline{\overline{V'} \cap V_0} \subset \overline{\Delta(V)}$.

Lemma 6. *Let* (D, W) *be a closed default theory. If* U *is* W-*complete, then the* U-*extensions of* (D, W) *are identical to its extensions.*

Proof. Consider any extension, that is, a set of the form $Th(V)$, where $\Sigma(V) = V$ (Proposition 2). By (11), $\overline{V} = V$; consequently, (12) and (13) give

$$Th(V) = Th(V \cap V_0) \tag{14}$$

and

$$\Sigma(V \cap V_0) = V.$$

From the last formula and Lemma 4, $\Delta(V \cap V_0) = V \cap V_0$, so that $V \cap V_0$ is a fixpoint of Δ, and $Th(V \cap V_0)$ is a U-extension. By (14), this set is identical to $Th(V)$.

Now consider any U-extension, that is, a set of the form $Th(V)$, where $\Delta(V) = V$. By (13) and Lemma 5,

$$\Sigma(\overline{V}) = \Sigma(V) = \overline{\Delta(V)} = \overline{V}.$$

Thus $\overline{V}$ is a fixpoint of Σ, so that $Th(\overline{V})$ is an extension (Proposition 2). By (12), this set is identical to $Th(V)$.

Proposition 4. *Let* (D, W) *be a closed default theory. If all models of* W *have the same finite cardinality, then the* F-*consequences of* (D, W) *are identical to its consequences.*

Proof. Let the common cardinality of all models of W be n. If the cardinality of U is n, then, by Lemma 6, the U-extensions of the theory are the same as its extensions. If not, then the set of worlds is empty, $\emptyset$ is the only fixpoint of Δ, and the set of all sentences is the only U-extension. It follows that the intersection of all U-extensions coincides with the intersection of all extensions.

Proposition 5. *Let* (D, W) *be a closed default theory with at most countably many object, function and predicate constants. If all models of* W *are infinite, then the* F-*consequences of* (D, W) *are identical to its consequences.*

Proof. If U is infinite, then, by Lemma 6, the U-extensions of the theory are the same as its extensions. If U is finite, then the set of worlds is empty, $\emptyset$ is the only fixpoint of Δ, and the set of all sentences is the only U-extension. It follows that the intersection of all U-extensions coincides with the intersection of all extensions.

Proposition 6. *For any default theory with the set of defaults (7), every fixpoint of the corresponding operator Δ is an equivalence class of the set of α-maximal worlds relative to the relation $\alpha^M = \alpha^{M'}$. Conversely, each of these equivalence classes is a fixpoint of Δ.*

Proof. By the definition of α^M, for any set of worlds V and any tuple of names ξ,

$$\alpha(\xi) \in Th^*(V) \Leftrightarrow \xi \in \bigcap_{M \in V} \alpha^M,$$

$$\neg\alpha(\xi) \in Th^*(V) \Leftrightarrow \xi \notin \bigcup_{M \in V} \alpha^M.$$

Consequently, for a default theory with the set of defaults (7), (***) is equivalent to the condition:

For any tuple of names ξ, if $\xi \in \bigcup_{M \in V} \alpha^M$ then $\xi \in \bigcap_{M' \in V'} \alpha^{M'}$,

that is to say, to the inclusion

$$\bigcup_{M \in V} \alpha^M \subset \bigcap_{M' \in V'} \alpha^{M'}.$$

Since $\Delta(V)$ is the largest set of worlds V' satisfying this condition,

$$\Delta(V) = \{M' : \bigcup_{M \in V} \alpha^M \subset \alpha^{M'}\}. \tag{15}$$

Assume that V is an equivalence class of the set of α-maximal worlds relative to the relation $\alpha^M = \alpha^{M'}$. Then, for some world M_0,

$$V = \{M : \alpha^{M_0} = \alpha^M\} = \{M : \alpha^{M_0} \subset \alpha^M\}.$$

The set union in (15) coincides with α^{M_0}, so that

$$\Delta(V) = \{M' : \alpha^{M_0} \subset \alpha^{M'}\} = V.$$

Take now any fixpoint V of Δ. If M' is a world from V, then, by (15),

$$\bigcup_{M \in V} \alpha^M \subset \alpha^{M'},$$

so that, for any $M \in V$, $\alpha^M \subset \alpha^{M'}$. We established this inclusion for any pair of worlds $M, M' \in V$, which means that, for every such pair, $\alpha^M = \alpha^{M'}$. Furthermore, V is nonempty, because, by (15), $\Delta(\emptyset)$ is the set of all worlds (which is assumed to be nonempty). For any world $M_0 \in V$,

$$V \subset \{M' : \alpha^{M_0} = \alpha^{M'}\},$$

and, by (15),

$$\Delta(V) = \{M' : \alpha^{M_0} \subset \alpha^{M'}\}.$$

Since V is a fixpoint of Δ, we conclude that

$$\{M' : \alpha^{M_0} \subset \alpha^{M'}\} \subset \{M' : \alpha^{M_0} = \alpha^{M'}\}.$$

This means that M_0 is α-maximal. Consequently, the fixpoint V is a subset of an equivalence class V_0 of α-maximal worlds. We know that V_0 is a fixpoint of Δ also; since Δ is monotone decreasing, it follows that $V = V_0$.

On Asking What a Database Knows

Raymond Reiter
Department of Computer Science
University of Toronto
Toronto, Canada M5S 1A4
email: reiter@ai.toronto.edu
and
The Canadian Institute for Advanced Research

Abstract

The by now standard perspective on databases, especially deductive databases, is that they can be specified by sets of first order sentences. As such, they can be said to be claims about the truths of some external world; the database is a representation of that world.

Virtually all approaches to database query evaluation treat queries as first order formulas, usually with free variables whose bindings resulting from the evaluation phase define the answers to the query. Following Levesque [8, 9], we argue that, for greater expressiveness, queries should be formulas in an epistemic modal logic. Queries, in other words, should be permitted to address aspects of the external world as represented by the database, as well as aspects of the database itself, i.e. aspects of what the database knows about that external world. We shall also argue that integrity constraints are best viewed as sentences about what the database knows, not, as is usually the case, as first order sentences about the external world. On this view, integrity constraints are modal sentences and hence are formally identical to a strict subset of the permissible database queries. Integrity maintenance then becomes formally identical to query evaluation for a certain class of database queries.

We formalize these notions in Levesque's language *KFOPCE*, and define the concepts of an answer to a query and of a database satisfying its integrity constraints. We also show that Levesque's axiomatization of *KFOPCE* provides a suitable logic for reasoning about queries and integrity constraints. Next, we show how to do query evaluation and integrity maintenance for a restricted, but sizable class of queries/constraints. An interesting feature of this class of queries/constraints is that Prolog's negation as failure mechanism serves to reduce query evaluation to first order theorem proving. This provides independent confirmation that negation as failure is really an epistemic operator in disguise. Finally, we provide sufficient conditions for the completeness of this query evaluator.

1 Introduction

The by now conventional perspective on databases, especially deductive databases, is that they are sets of first order sentences. As such, they can be said to be claims about the truths of some *external* world; the database is a symbolic representation of that world.

Virtually all approaches to database query evaluation treat queries as first order formulas, usually with free variables whose bindings resulting from the evaluation phase define the answers to the query. The sole exception to this is the work of Levesque [8, 9] who argues in favour of a generalized query language drawn from an epistemic modal logic. Queries, in other words, should be permitted to address aspects of the external world as represented in the database, as well as aspects of the database itself, i.e. aspects of what the database *knows*. To take a simple example, suppose $DB = \{p \vee q\}$.

- Query: p i.e. is p true in the external world?

  Answer: unknown.

- Query: Kp i.e. do you know that p is true in the external world?

  Answer: no.

- Query: $Kp \vee K\neg p$ i.e. do you know *whether* p?

  Answer: no.

Levesque's modal logic (called *KFOPCE*) also distinguishes between known and unknown individuals in the database and thus accounts for "regular" database values as well as null values. For example, suppose DB is

$$\{Teach(John, Math), (\exists x)Teach(x, CS), Teach(Mary, Psych) \vee Teach(Sue, Psych)\}.$$

- Query: $Teach(Mary, CS)$.
  Answer: unknown.

- Query: $KTeach(Mary, CS)$.
  Answer: no.

- Query: $K\neg Teach(Mary, CS)$
  Answer: no.

- Query: $(\exists x)KTeach(John, x)$ i.e. is there a known course which John teaches?
  Answer: yes, *Math*.

- Query: $(\exists x)KTeach(x, CS)$ i.e. is there a known teacher for CS?
  Answer: no.

- Query: $K(\exists x)Teach(x, CS)$ i.e. is someone known to teach CS without that someone necessarily being a known individual?
  Answer: yes.

- Query: $(\exists x)Teach(x, Psych)$ i.e. does someone teach $Psych$?

  Answer: yes, $Mary$ or Sue.

- Query: $(\exists x)KTeach(x, Psych)$ i.e. is there a known teacher of $Psych$?

  Answer: no.

- Query: $(\exists x)Teach(x, Psych) \wedge \neg Teach(x, CS)$ i.e. is there anyone who teaches $Psych$ and not CS?

  Answer: unknown.

- Query: $(\exists x)Teach(x, Psych) \wedge \neg KTeach(x, CS)$ i.e. does anyone teach $Psych$ who is not known to teach CS?

  Answer: yes, $Mary$ or Sue.

Levesque proposes that a database is best viewed as a set of first order sentences, and that it be queried by sentences of the richer language *KFOPCE*. This paper pursues this suggestion in various directions. Our principal results are as follows:

1. We provide a conceptual analysis of integrity constraints concluding that, contrary to the prevailing view, constraints are not first order sentences but modal sentences of *KFOPCE*. Moreover, testing for constraint satisfaction is identical to query evaluation.

2. *KFOPCE* is a suitable logic for reasoning about database queries and constraints. In particular, this logic provides the foundations for optimizing both queries and constraints.

3. For a sizable class of database queries and constraints we provide a sound Prolog-style evaluator which relies solely on first order theorem proving.

4. We provide sufficient conditions for the completeness of this evaluator and characterize a class of databases and queries guaranteeing the soundness and completeness of query evaluation.

This is an abridged version of Reiter [18], which contains the proofs omitted here because of space limitations, as well as results on query evaluation under the closed world assumption.

2 The Semantics of *KFOPCE*

KFOPCE is a first-order modal language with equality and with a single modal operator K (for "know"), constructed in the usual way from a set of predicate symbols, a countably infinite set of variable symbols and a countably infinite set of symbols called *parameters*. Predicate symbols take variables and parameters as their arguments. Parameters can be thought of as constants. Their distinguishing feature is that they are pairwise distinct and they define the domain over which quantifiers range, i.e. the parameters represent a single universal domain of discourse. *FOPCE* is the language *KFOPCE* without the modal K.[1]

A database of information about a world will be specified by a set of *FOPCE* sentences. We consider how Levesque defines the result of querying such a database with a sentence of *KFOPCE*. This requires first specifying a semantics for *KFOPCE*. An *atomic sentence* (of *KFOPCE*) is any atom of the form $P(p_1, \ldots, p_n)$, where P is an n-ary predicate symbol and $p_1, \ldots, p_n$ are parameters. A *world* is any set of atomic sentences that includes $p = p$ for each parameter p, and that does not include $p_1 = p_2$ for different parameters p_1 and p_2. The effect of this requirement on the equality predicate is that semantically the parameters are all pairwise distinct. A world is understood to be a set of true atomic sentences. The truth value of a sentence of *KFOPCE* in a world W and a set of worlds $\mathcal{S}$ is defined as follows:

1. If π is an atomic sentence, π is true in W and $\mathcal{S}$ iff $\pi \in W$.

2. $\neg w$ is true in W and $\mathcal{S}$ iff w is not true in W and $\mathcal{S}$.

3. $w_1 \wedge w_2$ is true in W and $\mathcal{S}$ iff w_1 and w_2 are both true in W and $\mathcal{S}$.

4. $(\forall x)w$ is true in W and $\mathcal{S}$ iff for every parameter p, $w|_p^x$ is true in W and $\mathcal{S}$.[2]

5. Kw is true in W and $\mathcal{S}$ iff for every $S \in \mathcal{S}$, w is true in S and $\mathcal{S}$.

Notice that condition 4 implies that, insofar as *KFOPCE* is concerned, the parameters constitute a single universal domain of discourse. The parameters are used to identify the known individuals. Notice also that when f is a *FOPCE* sentence (so that condition 5 need never be invoked in the truth recursion for f) then the truth value of f in W and $\mathcal{S}$ is independent of $\mathcal{S}$, and we can speak of the truth value of f in W alone.

When W is a world and $\mathcal{S}$ a set of worlds, the pair $(W, \mathcal{S})$ is a *model* of a set of *KFOPCE* sentences iff each sentence is true in W and $\mathcal{S}$. When Σ is a set of *KFOPCE* sentences, we write $\Sigma \models_{KFOPCE} w$ whenever the *KFOPCE* sentence w is true in all models of Σ. Similarly, when Σ is a set of *FOPCE* sentences and w is a *FOPCE* sentence, we write $\Sigma \models_{FOPCE} w$ with the obvious meaning.

Definition: Answer to a Query

Let Σ be a set of first order sentences, and q a *KFOPCE* formula with free variables $\vec{x}$. Let $\mathcal{M}(\Sigma)$ be the set of models of Σ. A tuple $\vec{p}$ of parameters is an *answer* to q (*wrt* Σ) iff for all $W \in \mathcal{M}(\Sigma)$, $q|_{\vec{p}}^{\vec{x}}$ is true in W and $\mathcal{M}(\Sigma)$.

Henceforth, we shall write $\Sigma \approx \sigma$ whenever the *KFOPCE* sentence σ is true in W and $\mathcal{M}(\Sigma)$ for all $W \in \mathcal{M}(\Sigma)$. In this notation, $\vec{p}$ is an answer to q iff $\Sigma \approx q|_{\vec{p}}^{\vec{x}}$. In particular, when q is a sentence (so that its answer should be *yes, no,* or *unknown*), then $\Sigma \approx q$ means *yes,* $\Sigma \approx \neg q$ means *no,* and neither means *unknown*.

---

[1] Notice that our language does not provide function or constant symbols. While the parameters have many features in common with constants, they have different semantics. In [9], Levesque generalizes the languages *FOPCE* and *KFOPCE* to include function symbols. In this, our preliminary cut at a theory of database queries and integrity constraints, we restrict ourselves to the function-free case.

[2] $w|_p^x$ is the result of substituting the parameter p for all free occurrences of the variable x in w.

Notice that $\Sigma \models_{KFOPCE} w$ implies $\Sigma \approx\!\!\!| w$, but not conversely.
Notice also that when σ is a *FOPCE* sentence,

$$\Sigma \approx\!\!\!| \ K\sigma \ \textit{iff} \ \Sigma \models_{FOPCE} \sigma$$

and

$$\Sigma \approx\!\!\!| \ \neg K\sigma \ \textit{iff} \ \Sigma \not\models_{FOPCE} \sigma.$$

Thus, under the relation $\approx\!\!\!|$, K acts like a first order provability relation and may be viewed as a formalization of the provability predicate of Bowen and Kowalski [1]. (See also Sadri and Kowalski [19].) The principal advantages of the modal approach to such a predicate are:

1. Formulas need not be reified.

2. Provability is formally specified. This includes the meaning of iterated modalities. Moreover, for formulas which "quantify into" modal contexts, the range of the quantified variables is precisely defined; they range over the parameters.

While we agree with Kowalski [7] that meta-level concepts like provability are extremely important for databases and AI in general - witness this paper - we believe that they require more precise semantic analyses than they have thus far been given. For this reason, Levesque's account of a provability relation seems to us an important step in the right direction.

It is also important to note that commitment to a modal *semantics* need not imply a concomitant commitment to a modal *proof theory*.[3] Indeed, most of this paper is devoted to showing how our modal semantics leads to first order - indeed Prolog-like - query evaluators.

3   What Is an Integrity Constraint?

The concept of an integrity constraint arises in databases, and in AI knowledge representation languages. The basic idea is that only certain database states are considered acceptable, and an integrity constraint is meant to enforce these legal states.

Integrity constraints have two flavours - static and dynamic. The enforcement of a static constraint depends only on the current state of the database, independently of any of its prior states. The fact that every employee must have a social security number is an example of a static constraint. Dynamic constraints depend on two or more database states. For example, if employee salaries must never decrease, then in no future database state may an employee's salary be less than it is in the current state. In this paper, we shall be concerned only with static integrity constraints.

The common perspective on integrity constraints is that they are first order sentences (e.g. Lloyd and Topor [11], Nicolas and Yazdanian [15], Reiter [16]). Corresponding to this perspective, there are at least four definitions in the literature of a database *DB* satisfying an integrity constraint *IC*:

---

[3]Although such a proof theory might have its uses. See Section 4

Definition 1 Consistency: open databases. (e.g. Kowalski [6])

$$DB \text{ satisfies } IC \text{ iff } DB + IC \text{ is satisfiable.}$$

Definition 2 Entailment: open databases. (e.g. Reiter [16])

$$DB \text{ satisfies } IC \text{ iff } DB \models IC.$$

Definition 3 Consistency: closed Prolog-like databases. (e.g. Sadri and Kowalski [19])

$$DB \text{ satisfies } IC \text{ iff } Comp(DB) + IC \text{ is satisfiable.}$$

Here, $Comp(DB)$ is the *completion* of DB in the sense of Clark [4]. Hence this notion is specific to Prolog-like databases, for which the completion is defined. It would not apply, for example, to databases with existentially quantified or disjunctive information.

Definition 4 Entailment: closed Prolog-like databases. (e.g. Lloyd and Topor [11])

$$DB \text{ satisfies } IC \text{ iff } Comp(DB) \models IC.$$

As already remarked, definitions 3 and 4 are peculiar to Prolog-like databases and do not have general applicability.[4] Our concern in this paper is with the most general possible notion of a database, without prior bias in favour of any kind of closed world assumption. Definitions 1 and 2 presume to address this general setting. Alas, neither of these two definitions correctly capture our intuitions. Consider the constraint about employees and their social security numbers:

$$(\forall x)emp(x) \supset (\exists y)ss\#(x, y) \tag{1}$$

1. Suppose $DB = \{emp(Mary)\}$. Then $DB + IC$ is satisfiable. But intuitively, the constraint should require DB to contain a social insurance number entry for $Mary$, so we want IC to be violated. Thus Definition 1 fails to capture our intuitions.

2. Suppose $DB = \{\ \}$. Intuitively, this should satisfy IC, but $KB \not\models IC$. So Definition 2 is inappropriate.

An alternative definition comes to mind when one sees that constraints like (1) intuitively are interpreted as statements not about the world but about the *contents* of the database, or about what it *knows*. Thus, (1) is attempting to say something like: Every employee *known* to the database must have a social security number, also *known* to the database. On this reading, (1) should be rendered by:

$$(\forall x)K emp(x) \supset (\exists y)K ss\#(x, y).$$

---

[4]Even in this case, there is no agreement on which of definitions 3 and 4 is appropriate. They are not equivalent.

Other Examples

1. To prevent a database from simultaneously assigning the properties *male* and *female* to the same individual, use the constraint

$$(\forall x)\neg(K(male(x) \land female(x))).$$

2. To force a database to assign one of the properties *male* and *female* to each individual, use the constraint

$$(\forall x)Kperson(x) \supset Kmale(x) \lor Kfemale(x).$$

3. To require that known instances of the relation *mother* have first argument a *female person* and second argument a *person*, use the constraint

$$(\forall x,y)Kmother(x,y) \supset K(person(x) \land female(x) \land person(y)).$$

4. To require that every known employee have a social security number, without necessarily knowing what that number is, use

$$(\forall x)Kemp(x) \supset K(\exists y)ss\#(x,y).$$

5. Functional dependencies in relational database theory are integrity constraints of a particular form. On our notion of a constraint, the functional dependency that social security numbers be unique would be represented by:

$$(\forall x,y,z)Kss\#(x,y) \land Kss\#(x,z) \supset Ky = z.$$

Many other kinds of dependencies have been investigated for relational databases. Most of these can be represented as first order sentences (Fagin [5], Nicolas and Gallaire [14]). The corresponding modalized forms of these first order sentences provide the correct reading of these dependencies, at least on our account of integrity constraints.

The view that integrity constraints are statements about the *content* of a database also serves to clarify a certain confusion in the literature about the different roles played by constraints and database formulas. According to the conventional account, constraints and databases are both first order sentences. Since constraints are external to the database, they do not enter into the query evaluation process. Yet, as first order sentences, they must express truths about the world, no less so than the database itself. Why then should they not contribute to answering queries? There is no clear answer to this in the literature. Nicolas and Gallaire [14] propose various pragmatic criteria for treating a formula as a constraint rather than as a component of the database, but there appear to be no general principles. On our account, no such principles are necessary. Truths about the world, namely first order sentences, belong in the database. Truths about the database, namely modalized sentences, function as integrity constraints.

We are finally lead to the following

Definition: Integrity Constraint Satisfaction

When IC is a $KFOPCE$ sentence, and Σ is a $FOPCE$ theory, Σ *satisfies* the integrity constraint IC iff $\Sigma \mathrel{\vDash\mkern-14mu\sim} IC$.

This notion of an integrity constraint and its satisfaction was first introduced by Reiter [17]. A related idea, appealing to a metatheoretic provability predicate in the context of Prolog databases, was introduced by Sadri and Kowalski [19].

Testing constraint satisfaction is identical to querying a first order database with a $KFOPCE$ sentence. Formally, then, in testing constraint satisfaction and in evaluating queries, we are faced with the same problem, namely determining whether $\Sigma \mathrel{\vDash\mkern-14mu\sim} w$ for Σ a $FOPCE$ theory and w a $KFOPCE$ formula. Most of this paper is about ways of doing this for certain formulas w and theories Σ.

4 Reasoning about Queries and Constraints

A natural question is this: What logic is appropriate for reasoning about queries and constraints? There are several reasons such a logic would be desirable. We might wish to determine whether certain constraints are redundant i.e. are entailed by the others. If a query or constraint were unsatisfiable, it would be important to know that. We might wish to optimize a given query or set of constraints in various ways, or explore their consequences.

The following is a simple consequence of the definitions of Section 2.

Theorem 1 *Suppose Σ is a FOPCE theory and α and β are KFOPCE sentences. Then $\Sigma \mathrel{\vDash\mkern-14mu\sim} \alpha$ and $\alpha \models_{KFOPCE} \beta$ implies $\Sigma \mathrel{\vDash\mkern-14mu\sim} \beta$.*

Corollary 1 (Equivalence of integrity constraints)
If IC and IC' are KFOPCE sentences and $\models_{KFOPCE} IC \equiv IC'$, then Σ satisfies IC iff Σ satisfies IC'.

This corollary provides a basis for integrity constraint simplification and optimization. If IC is an integrity constraint for Σ[5] and if IC' is equivalent to IC in the logic $KFOPCE$, then IC' may be used in the place of IC for maintaining the integrity of Σ. Presumably, we would want to do this whenever IC' is superior to IC in some way, for example computationally more feasible. Whatever the criteria might be for preferring IC', $KFOPCE$ provides a logic for proving equivalence and for transforming IC to IC'.

Corollary 2 (Equivalence of queries)
Suppose that q and q' are two KFOPCE queries with the same free variables $\vec{x}$, and IC is an integrity constraint for Σ. Suppose further that Σ satisfies IC, and that

$$IC \models_{KFOPCE} (\forall \vec{x})q \equiv q'.$$

Then q and q' are equivalent queries in the sense that the answers to each of them are the same: For all parameters $\vec{p}$, $\Sigma \mathrel{\vDash\mkern-14mu\sim} q|_{\vec{p}}^{\vec{x}}$ iff $\Sigma \mathrel{\vDash\mkern-14mu\sim} q'|_{\vec{p}}^{\vec{x}}$.

---

[5] IC might be a conjunction of all the constraints for Σ.

This provides a basis for query optimization. If the database Σ satisfies its constraints, and if, using these constraints, we can transform (using the logic *KFOPCE*) query q to an equivalent (within *KFOPCE*) query q', then the answers for q' are the same as for q. Chakravarthy, Grant and Minker [2] provides results on first order query optimization using first order constraints. Corollary 2 provides the formal foundations for such an analysis in the case of *KFOPCE* queries and constraints.

Levesque [8] provides a sound and complete axiomatization for *KFOPCE*. Since our concern in this paper is with theoretical foundations for queries and integrity constraints, we omit a description of Levesque's axiomatization. It is sufficient for our purposes to know that a suitable proof theory exists for reasoning about queries and constraints.

5   An Evaluator for a Class of Queries

Our objective in this section is to establish a soundness result for a class of Prolog-like queries and constraints. Unlike Prolog queries, explicit existential quantifiers will be allowed, as will the modal K. Our evaluator will rely on negation-as-failure, and left-to-right evaluation of queries. As is the case with Prolog, negation-as-failure must be avoided on subgoals mentioning unbound variables. The following definition guarantees this (as we shall see), and is a suitable generalization for our query language of the notion of a safe-for-negation query (Lloyd [12]).

Definition

The *safe KFOPCE* formulas are the smallest set such that

1. A first order formula (having any syntactic form whatever) is safe.

2. If σ is safe, so are:
   $K\sigma$,
   $(\exists v)\sigma$, where v is any variable.
   $\neg\sigma$, whenever σ is a sentence.

3. If σ_1 is safe and $\vec{x}$ are all the free variables of σ_1, then $\sigma_1 \wedge \sigma_2$ is safe whenever $\sigma_2|_{\vec{p}}^{\vec{x}}$ is safe for all parameters $\vec{p}$.

Examples

- The following are safe:
  $p(x,y) \wedge Kq(x) \wedge Kr(x)$.
  $(\exists x)\neg r(x)$.
  $\neg K\neg[(\forall x,y)p(x,y) \supset q(x) \vee r(x)]$.
  $p(x,y) \wedge \neg Kq(x) \wedge \neg Kr(y)$.
  $(\exists x,y)p(x,y) \wedge \neg K[q(x) \vee \neg r(y)]$.

- The following are not safe:
  $(\exists x)\neg Kr(x)$.
  $t(x) \wedge \neg Kq(x) \wedge \neg Kr(y)$.
  $\neg Kq(x) \wedge Kr(x)$.

Definition

The *subjective KFOPCE* formulas are the smallest set such that

1. $t_1 = t_2$ is subjective for arbitrary terms t_1 and t_2.

2. Kf is subjective whenever f is first order.

3. If π is subjective, so are $K\pi$, $(\exists x)\pi$, $\neg\pi$.

4. If π_1 and π_2 are subjective, so are $\pi_1 \wedge \pi_2$.

The subjective formulas say nothing about the external world; they address only the epistemic state of a database.

The following is the key definition of this section. As we shall see, it provides a class of queries for which a Prolog-style query evaluator using negation-as-failure can be proved sound.

Definition

A *KFOPCE* formula is *admissible* iff

1. It is safe.

2. Its quantified variables are distinct from one another, and from its free variables.

3. The scope of every existential quantifier is a subjective formula, or is first order.

4. The scope of every negation sign is a subjective formula, or is first order.

Examples

1. All but the last of the example queries of Section 1 are admissible. The following is not admissible:

$$(\exists x)\neg KTeach(x, CS) \wedge KTeach(x, Psych)$$

2. The example integrity constraints of Section 3 have the following equivalent representations as admissible sentences:

$$\neg(\exists x)Kemp(x) \wedge \neg(\exists y)Kss\#(x, y)$$

$$\neg(\exists x)K(male(x) \wedge female(x))$$

$$\neg(\exists x)Kperson(x) \wedge \neg Kmale(x) \wedge \neg Kfemale(x)$$

$$\neg(\exists x, y)Kmother(x, y) \wedge \neg K(person(x) \wedge female(x) \wedge person(y))$$

$$\neg(\exists x)Kemp(x) \wedge \neg K(\exists y)ss\#(x, y)$$

$$\neg(\exists x, y, z)Kss\#(x, y) \wedge Kss\#(x, z) \wedge \neg Ky = z$$

Notice the close similarity between the transformations made of the original constraints to the above admissible formulas and the transformations on arbitrary first order queries introduced by Lloyd and Topor [11] for conversion to Prolog executable form.

3. $p(x) \wedge K q(x)$ is admissible.
$(\exists x) p(x) \wedge K q(x)$ is not.

5.1   A Sound Evaluator for Admissible Queries

Consider the following Prolog code for a meta-interpreter *demo*:

$$
\begin{aligned}
demo(f, \Sigma) &\leftarrow first\text{-}order(f), prove(f, \Sigma). \\
demo(\neg w, \Sigma) &\leftarrow modal(w), not\, demo(w, \Sigma). \\
demo(K w, \Sigma) &\leftarrow demo(w, \Sigma). \\
demo((\exists x) w), \Sigma) &\leftarrow modal(w), demo(w, \Sigma). \\
demo(w_1 \wedge w_2, \Sigma) &\leftarrow modal(w_1 \wedge w_2), demo(w_1, \Sigma), demo(w_2, \Sigma).
\end{aligned}
$$

As usual, the clausal antecedents are executed from left to right and *not* executes as negation-as-failure. The auxiliary predicates *first-order(f)* and *modal(w)* are true when f is a *FOPCE* formula, and w is a modal formula i.e. mentions K at least once. We shall assume that the predicate *prove* is a first order theorem prover with the following properties:

Let π be an enumeration of all those parameters $\vec{p}$ such that $\Sigma \models_{FOPCE} f_{\vec{p}}^{\vec{x}}$, where $\vec{x}$ are all the free variables of f. The first time $prove(f, \Sigma)$ is called, it binds $\vec{x}$ to the first $\vec{p}$ in the enumeration π, and the call succeeds. If there is no first $\vec{p}$ in the enumeration π, the call $prove(f, \Sigma)$ fails. The second time $prove(f, \Sigma)$ is called (which can happen only if the first call succeeds and a subsequent call to *demo* fails), it binds $\vec{x}$ to the second $\vec{p}$ in the enumeration π, and the call succeeds. If there is no second $\vec{p}$ in the enumeration π, the call $prove(f, \Sigma)$ fails. And so on. Thus, successive calls to *prove* iterate through the enumeration π, binding $\vec{x}$ to the next $\vec{p}$ in the enumeration, failing only if π runs out of tuples.[6] When π is infinite it is possible for *demo* not to terminate.

Notice that we make no assumption about how *prove* goes about its business. It can be realized by any sound and complete first order theorem prover. Moreover - and this is more important - no assumption is made about the first order theory Σ, not even that it is finite. It can be an open database, or closed, or clopen. It could, for example, be a Datalog program and *prove* could be realized using negation-as-failure. None of this is of any concern to the meta-interpreter *demo*; it is totally decoupled from the form of the database Σ and the workings of the first order theorem prover *prove*.

Theorem 2 (Soundness of *demo***)** *Suppose that w is admissible, that w has free variables $\vec{x}$, and that Σ is a satisfiable FOPCE theory.*

1. *If $demo(w, \Sigma)$ succeeds, then the variables $\vec{x}$ are all bound to parameters $\vec{p}$, and $\Sigma \models w|_{\vec{p}}^{\vec{x}}$.*

2. *If $demo(w, \Sigma)$ finitely fails, then for all parameters $\vec{p}$, $\Sigma \not\models w|_{\vec{p}}^{\vec{x}}$.*

---

[6]Notice that this assumption implies that *prove* is both sound and complete.

Theorem 2 is a *relative* soundness result; it is contingent on the availability of a sound and complete first order theorem prover *prove*, capable of enumerating all $\vec{p}$ for which, whenever f is first order, $\Sigma \models_{FOPCE} f|_{\vec{p}}^{\vec{x}}$. In a sense, Theorem 2 is not novel. Levesque [8, 9] proves that *all KFOPCE* queries may be soundly and completely evaluated using only first order theorem proving, although his method suffers serious computational problems. The novelty of Theorem 2 is the provision of a Prolog-like evaluator well suited to the database setting.

5.2   Queries Revisited

Normally for deductive databases queries are taken to be conjunctions of first order literals. The answers to such a query are the bindings of its free variables, resulting from a successful proof of the query, usually under a Prolog-style evalutor. We now consider the analogous class of KFOPCE queries and our meta-evaluator *demo* on these. Define a *normal* query to be any *KFOPCE* formula of the form $L_1 \wedge \cdots \wedge L_n$ where L_i is a first order literal, or has the form Kl or $\neg Kl$ where l is a first order literal.

It is easy to see that a normal query is admissible iff it is safe. Hence, by Theorem 2, *demo* soundly evaluates all safe normal queries. Since safety is the *KFOPCE* version of Prolog's safe-for-negation requirement, this means that for normal queries, *demo* is as close to a Prolog-like query evaluator as one could possibly hope for. So it seems that generalizing conjunctive first order queries to normal *KFOPCE* queries genuinely increases the expressiveness of the query language *without sacrificing the computational advantages of first order query evaluation.*

5.3   Integrity Constraints Revisited

Recall the intuitions about integrity constraints that lead to their formalization as *KFOPCE* sentences: they are statements not about the world, but about the contents of the database itself. On this intuition, integrity constraints can be expected to assume a particular syntactic form: every *FOPCE* formula mentioned by a constraint should occur within the scope of a K operator. In other words, integrity constraints are subjective *KFOPCE* sentences. Moreover, there seems to be no need for iterated modalities in representing constraints. If we call formulas without iterated modalities K_1 formulas, then we are lead to the conjecture that integrity constraints are naturally represented by pure K_1 *KFOPCE* sentences. All the examples of Section 3 are pure K_1 sentences. Such sentences are of interest for the following reason:

Result 1 *If σ is a subjective K_1 sentence, then σ is admissible iff it is safe and its quantified variables are distinct.*

So *demo* soundly evaluates *all* integrity constraints provided they are safe and sensibly quantified. Since safety is the *KFOPCE* version of Prolog's safe-for-negation requirement, this means that for constraints, *demo* is as close to a Prolog-like query evaluator as one could possibly hope for.

Something like the (inconsequential) requirement that σ's quantified variables be distinct is necessary. To see why, consider *demo*'s execution on the sentence $(\exists x)[[(\exists x)Kp(x)] \wedge Kq(x)]$.

6   On the Completeness of *demo*

Section 5.1 provided a sound evaluator *demo* for admissible queries and constraints. Under what circumstances is *demo complete* for admissible formulas α, i.e. when is $demo(\alpha, \Sigma)$ guaranteed to return? Our purpose in this section is to provide some sufficient conditions for the completeness of *demo*.

6.1   On Queries with Finitely Many Answers

Definition

Suppose Σ is a set of *FOPCE* sentences, and w a *KFOPCE* formula with free variables $\vec{x}$. Define

$$Instances(w, \Sigma) = \{\vec{p} \mid \text{the } \vec{p} \text{ are parameters and } \Sigma \mathrel{\rlap{\hspace{0.1em}\sim}\approx} w|_{\vec{p}}^{\vec{x}}\}.$$

Notice that when w is first order,

$$Instances(w, \Sigma) = \{\vec{p} \mid \text{the } \vec{p} \text{ are parameters and } \Sigma \models_{FOPCE} w|_{\vec{p}}^{\vec{x}}\}.$$

Definitions

Let $\mathcal{F}_\Sigma$ be any set of *FOPCE* formulas with the property that whenever $f \in \mathcal{F}_\Sigma$, then $Instances(f, \Sigma)$ is finite.

Note that $\mathcal{F}_\Sigma$ need *not* be the set of all *FOPCE* formulas with finitely many instances. Although it could be so chosen, it could also be a proper subset of these.

The *KFOPCE* formulas which are *almost admissible (a.a.) wrt* $\mathcal{F}_\Sigma$ are the smallest set such that

1. If $f \in \mathcal{F}_\Sigma$, then f is a.a. wrt $\mathcal{F}_\Sigma$.

2. If σ is a subjective a.a. sentence wrt $\mathcal{F}_\Sigma$, then $\neg\sigma$ is a.a. wrt $\mathcal{F}_\Sigma$.

3. If σ is a subjective a.a. formula wrt $\mathcal{F}_\Sigma$, then $(\exists x)\sigma$ is a.a. wrt $\mathcal{F}_\Sigma$.

4. If σ is an a.a. formula wrt $\mathcal{F}_\Sigma$, so is $K\sigma$.

5. If σ_1, with free variables $\vec{x}$, is an a.a. formula wrt $\mathcal{F}_\Sigma$, and σ_2 is a *KFOPCE* formula such that $\sigma_2|_{\vec{p}}^{\vec{x}}$ is an a.a. formula wrt $\mathcal{F}_\Sigma$, then so is $\sigma_1 \wedge \sigma_2$.

Remark 1 *Every a.a. formula wrt $\mathcal{F}_\Sigma$ is safe.*

Remark 2 *If α is an a.a. formula wrt $\mathcal{F}_\Sigma$, and if the quantified variables of α are distinct from one another and from the free variables of α, then α is admissible.*

We shall call any formula satisfying the condition on variables of the previous remark *admissible wrt* $\mathcal{F}_\Sigma$.

As an immediate corollary of Theorem 2, we have:

Corollary 3 *demo soundly evaluates all formulas admissible wrt $\mathcal{F}_\Sigma$.*

Lemma 1 *Whenever w is admissible wrt $\mathcal{F}_\Sigma$, Instances(w, Σ) is finite.*

Our principal result is the following:

Theorem 3 (Completeness of *demo* for formulas admissible wrt $\mathcal{F}_\Sigma$) *If w is admissible wrt $\mathcal{F}_\Sigma$, then demo(w, Σ) returns.*

6.1.1 Recovering All Answers to Queries

Our soundness and completeness results for queries q admissible wrt $\mathcal{F}_\Sigma$ (Corollary 3 and Theorem 3) only guarantee that just one answer for q (if there are any at all) will be returned by *demo*. By Lemma 1, we know that q has finitely many answers. By forcing failure on q just after *demo* succeeds, are we guaranteed to iterate through all these finitely many answers to q? In other words, does *demo* allow us to recover all answers to q?

Assume that $\mathcal{F}_\Sigma$ contains at least one formula which always finitely fails, say $p_1 = p_2$ for distinct parameters p_1 and p_2. Then $q \wedge (p_1 = p_2)$ is admissible wrt $\mathcal{F}_\Sigma$ whenever q is. By Theorem 3, $demo(q \wedge (p_1 = p_2), \Sigma)$ must finitely fail. Let $\vec{x}$ be the free variables of q. We wish to know whether the bindings for $\vec{x}$ iterate through all the (finitely many) answers to q during the failed execution of $demo(q \wedge (p_1 = p_2), \Sigma)$. To see that this must be so, assume to the contrary that some answer $\vec{a}$ to q is missed in this iteration. Then $demo(q \wedge (\vec{x} = \vec{a}), \Sigma)$ must finitely fail.[7] But $\Sigma \not\approx (q \wedge (\vec{x} = \vec{a}))|_{\vec{a}}^{\vec{x}}$ so by soundness, this call to *demo* must succeed.

We have shown that, whenever q with free variables $\vec{x}$ is admissible wrt $\mathcal{F}_\Sigma$, the Prolog call

$$\leftarrow demo(q, \Sigma), write(\vec{x}), nl, fail.$$

will print all answers to q (possibly with repetitions, of course).

6.2 Completeness of demo for Elementary Databases

By relying on the results of the previous section we now define a sizable class of databases and queries for which *demo* can be proved complete.

Definitions

The *positive existential* (p.e.) FOPCE formulas are defined by the smallest set such that:

1. An atomic formula other than an equality atom is p.e.

2. If w is p.e., so is $(\exists x)w$.

3. If w_1 and w_2 are p.e. so are $w_1 \wedge w_2$ and $w_1 \vee w_2$.

A *rule* is a sentence of the form $(\forall \vec{x})A \supset B$ where A is a conjunction of atomic formulas other than equality atoms, B is a p.e. formula, and every variable of $\vec{x}$ occurs free in A.

A first order theory is *elementary* iff it is a set of p.e. sentences and rules. Notice that elementary theories make no mention of equality.

---

[7] Strictly speaking, we are assuming here that $\mathcal{F}_\Sigma$ contains the atoms $x = p$ for all variables x and parameters p, not an unreasonable assumption.

Elementary databases are analogous to the deductive databases widely studied in the logic programming community (e.g. Lloyd and Topor [11]). They are more general than deductive databases by admitting disjunctions and existential quantification. They are slightly less general by requiring that for rules $(\forall \vec{x})A \supset B$, the variables of $\vec{x}$ must all occur free in A, although this is a minor restriction in practice.

Definition

Suppose w is a FOPCE formula with free variables $\vec{x}$. Then w has *disjunctively linked variables* iff for each of its subformulas of the form $w_1 \vee w_2$, those free variables of w_1 which are among $\vec{x}$ are precisely the same as those of w_2 which are among $\vec{x}$.

Example

- The following have disjunctively linked variables:
  $P(a,b) \vee Q(a,c)$
  $(\forall x)(U(x) \vee W(x))$
  $P(a,x) \vee Q(x,x)$
  $(\exists y,z)(P(y,x) \vee R(y,z,x)) \vee (\exists u)(P(u,a) \wedge Q(u,x))$

- The following do not:
  $(\forall x)(U(x) \vee W(y))$
  $P(x,y) \vee Q(y,z)$

The following is the principal result of this section:

Theorem 4 *Suppose Σ is an elementary theory mentioning only finitely many distinct parameters. Let*

$$
\begin{aligned}
\mathcal{F}_\Sigma \;=\; & \{\pi \mid \pi \text{ is a p.e. formula with disjunctively linked variables}\} \;\cup \\
& \{p = p' \mid p \text{ and } p' \text{ are parameters}\} \;\cup \\
& \{p \neq p' \mid p \text{ and } p' \text{ are parameters}\} \;\cup \\
& \{x = p, p = x \mid x \text{ is a variable and } p \text{ is a parameter}\}.
\end{aligned}
$$

Then demo is a sound and complete evaluator for all queries admissible wrt $\mathcal{F}_\Sigma$.

7 Summary and Discussion

Following Levesque, we have argued in favour of a generalized query language for databases, a language which addresses the database's state of knowledge about the external world, as well as truths about that world. We have also argued that integrity constraints are statements about the contents of a database, not about the world. They are thus metatheoretic in character. We have appealed to Levesque's logics *FOPCE* and *KFOPCE* to formalize the concepts of an answer to a query and of a database satisfying its constraints. The logic *KFOPCE* also turns out to be suitable for reasoning about queries and constraints, and an appropriate vehicle for query and constraint optimization. For a sizable class of queries - the so-called admissible queries - we have provided a sound Prolog-like evaluator. Finally, we have considered some conditions under which this evaluator is complete.

There are many issues which we have not explored, but which deserve attention:

1. Epistemic query languages provide very subtle distinctions. Differences between truth and knowledge, and known and unknown individuals place a considerable burden on users to make their intentions clear in formulating queries and constraints. For naive users, this probably demands too much. There is a need for suitable interfaces which can elicit from users the precise nature of their requests for translation into the appropriate *KFOPCE* queries. This is a considerably more complex problem than the (already difficult) task of providing first order query interfaces to naive users. So far as we know, there has been no work done along these lines.

2. Our results should be extended to Levesque's [9] languages $\mathcal{L}$ and $\mathcal{KL}$ which admit function symbols. Among other things, this will allow databases to address null values properly.

3. Notice that we appealed to the relation $|\approx$ to define the concepts of constraint satisfaction and answer to a query. Now $\approx$ is an extra-logical notion, so constraint satisfaction and query evaluation are not defined *within KFOPCE* i.e. this concept is not defined in terms of *KFOPCE validity*. Recently, Levesque [10] has defined a logic within which one can define this concept. Briefly, this logic has two modalities: O (for *only know*) and K (for *know*). We can then say that KB satisfies IC iff $OKB \supset KIC$ is a valid sentence of this logic. The consequences for database theory of this notion remain to be explored.

4. We have not explored mechanisms for incremental modifications to a knowledge base. Usually a knowledge base will be known to satisfy its constraints. When a (normally) small change is made to it, it should not be necessary to verify all its constraints all over again. Rather, only enough computation should be devoted to verify the change in its state, given that its prior state was acceptable. Nicolas [13] provides such mechanisms for relational databases, as do Lloyd and Topor [11] for deductive databases. Similar mechanisms must be devised for our concept of integrity checking.

5. Many knowledge representation languages (e.g. Chung et al [3]) provide mechanisms for *procedural attachment* which are invoked whenever a change is made to the knowledge base state. Such procedures normally check to see whether certain conditions hold in the current state and if so, may change this state in various ways. Such changes may trigger other procedures, and so on. A simple example is a procedure triggered by an update of an employee record. It might then search for a social security entry for that employee and, failing in this, request this entry from the user. Clearly, this is a procedural version of the integrity constraint

$$(\forall x)Kemp(x) \supset (\exists y)Kss\#(x,y).$$

In general, there is an intimate connection between procedural attachment and integrity constraints. It would be worthwhile exploring this relationship, perhaps with two objectives in mind:

(a) Since there is a logic of integrity constraints, we can explore the consequences of the constraints, hence of their procedural incarnations.

(b) Correctness proofs should be possible for the procedures relative to their logically specified constraints.

Acknowledgements

Hector Levesque helped a lot on this one. Gerhard Lakemeyer provided valuable comments on an earlier draft. Michael Gelfond and Vladimir Lifschitz made a number of interesting observations on the role of autoepistemic logic in this setting, as well as in artificial intelligence in general. This research was supported by grant A9044 of the National Science and Engineering Council of Canada.

References

[1] K. Bowen and R. Kowalski. Amalgamating language and meta-language in logic programming. In K.L. Clark and S.A. Tarnlund, editors, *Logic Programming*, pages 153–172. Academic Press, New York, 1982.

[2] U.S. Chakravarthy, J. Grant, and J. Minker. Logic based approach to semantic query optimization. *ACM Transactions on Database Systems*, in press.

[3] L. Chung, D. Rios-Zertuche, B. Nixon, and J. Mylopoulos. Process management and assertion enforcement for a semantic data model. Technical report, Department of Computer Science, University of Toronto, 1987.

[4] K.L. Clark. Negation as failure. In H. Gallaire and J. Minker, editors, *Logic and Data Bases*, pages 292–322. Plenum Press, New York, 1978.

[5] R. Fagin. Horn clauses and database dependencies. In *ACM Symp. on Theory of Computing*, pages 123–134, 1980.

[6] R. Kowalski. Logic for data description. In H. Gallaire and J. Minker, editors, *Logic and Data Bases*, pages 77–103. Plenum Press, New York, 1978.

[7] R. Kowalski. *Logic for Problem Solving*. Elsevier Science Publishers B.V. (North-Holland), 1979.

[8] H. L. Levesque. *A Formal Treatment of Incomplete Knowledge Bases*. PhD thesis, Department of Computer Science, University of Toronto, 1981.

[9] H. L. Levesque. Foundations of a functional approach to knowledge representation. *Artificial Intelligence*, 23:155–212, 1984.

[10] H. L. Levesque. All I know: A study in autoepistemic logic. *Artificial Intelligence*, 2:263–309, 1990.

[11] J. W. Lloyd and R.W. Topor. A basis for deductive database systems. *Journal of Logic Programming*, 2:93–109, 1985.

[12] J.W. Lloyd. *Foundations of Logic Programming*. Springer Verlag, second edition, 1987.

[13] J.M. Nicolas. Logic for improving integrity checking in relational databases. *Acta Informatica*, 18(3):227–253, 1982.

[14] J.M. Nicolas and H. Gallaire. Data base: Theory vs. interpretation. In H. Gallaire and J. Minker, editors, *Logic and Data Bases*, pages 33–54. Plenum Press, New York, 1978.

[15] J.M. Nicolas and K. Yazdanian. Integrity checking in deductive data bases. In H. Gallaire and J. Minker, editors, *Logic and Data Bases*, pages 325–344. Plenum Press, New York, 1978.

[16] R. Reiter. Towards a logical reconstruction of relational database theory. In M.L. Brodie, J. Mylopoulos, and J.W. Schmidt, editors, *On Conceptual Modelling: Perspectives from Artificial Intelligence, Databases and Programming Languages*, pages 191–233. Springer, New York, 1984.

[17] R. Reiter. On integrity constraints. In M. Vardi, editor, *Proceedings of the Second Conference on Theoretical Aspects of Reasoning about Knowledge*, pages 97–111. Morgan Kaufmann Publishers, Inc., 1988.

[18] R. Reiter. What should a database know? Technical report, Department of Computer Science, University of Toronto, 1990.

[19] F. Sadri and R. Kowalski. An application of general purpose theorem-proving to database integrity. In J. Minker, editor, *Foundations of Deductive Databases*. Morgan Kaufmann Publishers, Inc., Palo Alto, California, 1987.

Two Kinds of Program Specifications

John McCarthy
Computer Science Department
Stanford University
Stanford, CA 94305

Almost all computer programs interact in some way with the world outside the computer, but this interaction is rather trivial for most programs whose specification or verification has been studied. Thus a program for computing a mathematical function generally has trivial interaction, whereas interaction with the outside world is the essence of a program for controlling the landings of airplanes.

In one case, the program is appropriately specified entirely in terms of the relation between its inputs and outputs. In the other case, this relation is important, but even more important is what the program accomplishes in the external world. To distinguish these cases, we borrow some jargon from the philosophy of speech acts and talk about *illocutionary* and *perlocutionary* specifications. You may also call the two kinds *input-output* specifications and *accomplishment* specifications, but I have been advised to retain the jargon in order to emphasize the distinctness of the ideas.

The philosophical terms come from the following distinction. Telling someone something is an illocutionary speech act; getting him to know it is perlocutionary. Likewise, commanding someone to do something is illocutionary; getting him to do it is perlocutionary. Thus illocutionary speech acts are defined in terms of what is said, and perlocutionary acts are defined in terms of what the saying accomplishes.

Let us return to the example of a program for controlling the landings of airplanes. Its illocutionary specifications include that it only outputs "Cleared to land." when the previous airplane appears to be off the runway. Here "appears" is taken in the sense that the program has received inputs, e.g. from a television camera, that it can interpret as meaning that the previous airplane is off the runway. The corresponding perlocutionary specification is that it outputs "Cleared to land." only when the previous airplane actually is off the runway.

Naturally, we would like to prove that the illocutionary specification implies the perlocutionary specification. Obviously, this can be done only on the basis of assumptions about the world, which, in this case, includes the television camera, its view of the runways, and its connection to the computer. Assumptions about the world are never absolutely guaranteed, but we often bet our lives and those of our families, friends and customers on such assumptions.

While the perlocutionary specifications are usually our ultimate interest, illocutionary specifications are often less problematical, because they do not involve assumptions about the outside world.

Using formal methods in proving perlocutionary specifications requires formalization of assumptions about the world. The tool for this is mathematical logic. In the example of a landing control program, there are at least the following kinds of assumptions.

1. Properties of objects located in 3-dimensional space.

2. Properties of motion.

3. Properties of observation.

4. Properties of communication.

5. Properties of obedience to instructions.

We will discuss how these assumptions can be expressed conveniently as logical sentences.

Exploration with *Mathematica*

Dana S. Scott
Hillman University Professor of
Computer Science, Mathematical Logic, and Philosophy
Carnegie Mellon University
Pittsburgh, Pennsylvania

During the Fall Semester of 1989 I presented a junior/senior mathematics course on **projective geometry** at my university, with a syllabus organized along quite traditional lines. I used a Macintosh II computer and the *Mathematica* symbolic computation program, a product of Wolfram Research, Inc. The lectures were delivered primarily from the console, using a screen attached to the computer and projected by an overhead projector. The students were expected to do their homework on a computer, and the final examination was a project on a topic selected individually by each student, again, to be done on the computer.

During the present Fall Semester of 1990 I am presenting a course on **methods of symbolic computation**. The audience is very mixed, and we are doing a selection of topics, some selected by the students and some selected by the instructor. An attempt is being made to tie in the individual work of the students with work for other courses. This year we have a classroom with a cluster of NeXT computers, with 12 for the students and one for the lecturer, and each machine runs *Mathematica*. There is a file server for distributing the class materials. A better projection device is being used this year giving a large image of what the instructor has on his screen. The cluster is open 24 hours a day, but it is reserved six hours a week for class work in two different classes. This is an experimental installation for testing teaching methods. Evaluation of student work again will be done by individual projects.

In my presentation, I will demonstrate on a computer several examples from these courses, and show how I use the machine in teaching. I will then discuss several lessons gained from the experience and suggestions I have for future developments.

Composition Operators
for Logic Theories

Antonio Brogi, Paolo Mancarella, Dino Pedreschi, Franco Turini

Dipartimento di Informatica, Università di Pisa
Corso Italia,40 — 56125 Pisa, Italy
{brogi,paolo,pedre,turini}@dipisa.di.unipi.it

Abstract

Some basic meta-level operators for putting logic theories together are introduced, which relate to set-theoretic union, intersection and difference. Both a transformational and an interpretive characterization of the operators are provided and proved equivalent. The former definition says how to syntactically construct a new theory out of two given theories, the latter provides a meta-level interpretation of the same operators. A declarative — both model-theoretic and fixpoint — semantics of the operators is also provided, allowing one to assign the minimal model of the resulting theory as a function of the models of the argument theories. Some examples from default reasoning, knowledge assimilation, inheritance networks and hypothetical reasoning are presented to demonstrate the expressive power of the operators.

1   Introduction

One of the most fascinating aspects of logic programming is its declarativeness. The property of being declarative is usually defined by stating that programs defines *what* has to be computed without saying too much about *how* the computation has to be carried on. Another way of measuring declarativeness is to see how close is the program to its abstract semantics. In fact, Horn Clause Logic has the appealing property that its model-theoretic, fixpoint and operational semantics are all equivalent [vEK76].

Several extensions of logic programming are based on a meta-level description of programs by means of systems of separate *theories,* in order to address issues like modularity, programming-in-the-large, incremental development of programs, meta-programming, dynamic theory evolution, forms of non-monotonic reasoning, inheritance and object-orientation.

The basic semantics of logic programming does not extend in a straightforward way to the granularity of theories and their compositions. As far as compositionality is concerned, the minimal (Herbrand) model semantics is too coarse. For instance, it is not possible to construct the minimal model of the union of two logic programs in terms of the minimal models of the separate programs. Different efforts have been devoted to define suitable semantic frameworks to cope with compositionality [LM84, MP88, GS89, MPRT90]. In [MP88,MPRT90] an algebra of logic programs is defined in terms of some basic meta-level operators acting on logic theories as a whole. The immediate consequence operator is taken to be the semantic counterpart of a theory, and the syntactic operators are mapped onto transformations over the semantics of the separate theories.

This algebraic approach leaves two main issues open. On one side, although the immediate consequence semantics for theory composition indirectly entails a model-theoretic characterization, it would be useful to provide a *direct* model-theoretic characterization, i.e. one which given the class of all the models of two theories produces the models of their compositions.

The second issue concerns an interpretive version of the composition operators: indeed, such operators are defined as syntactic transformations of the theories, and produce a new theory realizing the desired composition. It is then natural to investigate whether an interpreter exists which emulates the effect of the combined theory without actually building it.

The present paper is mainly devoted to address the above issues by introducing both a model-theoretic characterization of the composition operators and an interpretive one, based on the technique of the logic meta-interpreters. In due turn, the two approaches are proved to be equivalent each other, and with respect to the original algebraic characterization. As a consequence, a number of alternative ways of looking at the operators is now available: 1) operational / transformational, 2) operational / interpretive, 3) model-theoretic, 4) denotational (immediate consequence operator), 5) logic (connectives in the completed program).

We believe that a third way of measuring the declarativeness of a computation formalism is too see how easy is to define operations on program texts capable of realizing intended semantic transformations. More precisely, given a program text P with an abstract semantics [P] and an abstract semantics [P'] = f([P]), where f is a function from abstract semantics to abstract semantics, find a transformation T such that $[T(P)] = [P']$.

It should be apparent that the problem is more easily solvable if the program text P is very close to its abstract semantics [P]. The semantics driven definition of operators on logic theories enforce our claim that logic programming is definitely a declarative programming paradigm.

The availability of transformations which implement well defined semantic mappings has very interesting implications on how programs can be constructed in this framework. There are currently two main streams in approaching the problem of building programs in a better engineered fashion: one is the transformational approach, in which programs are built via successive transformations of initial specifications; the other one is the modular approach, in which programs are built by connecting together pre-existing modules, which are viewed as black boxes. Our approach combines both of them. Indeed our operators combine together existing modules in order to build new ones and, in doing that, transformations are applied to the existing modules.

Another viewpoint on our work is to consider the operators as the basic operations of a language for building programs, i.e. as the basic operations for a meta-language for a modular version of logic programming. Indeed, meta-programming is a typical feature of logic programming, but the meta-primitives which are usually provided work at the level of terms and clauses (e.g. clause, univ in Prolog systems) or, if they work at the level of programs (e.g. assert, retract), they are not semantically well defined and, in all cases, they handle one single module.

With this general motivations in mind, some evidence is provided in the paper to demonstrate the expressive power of the composition operators, using examples from default reasoning, knowledge assimilation and hierarchical reasoning. These examples are carried out emphasizing the modular / transformational approach to the construction of (knowledge-based) systems. These applications do not contribute to the fields of hierarchical reasoning and of default reasoning *per se*. They rather aim at putting well known results and problems into new perspectives, in which novel formal tools can be exploited, namely the various semantics of the operators. Our hope is that the new perspective and the new formal tools can hint in the future new or better solutions in the application field.

An outline of the paper follows. In Section 2, the union and intersection operators are introduced, while in Section 3 a model-theoretic definition is presented. Section 4 contains the meta-level definition of the theory-level operators. In Section 5, a retraction operator is defined.

2 Union and intersection of logic theories

Here we introduce two basic composition operators, union and intersection, following [MP88]. In the sequel, the term *theory* is a synonym of logic program, i.e. a set of definite Horn clauses. The definition of the immediate consequence operator (T_P) is the standard one, as in [vEK76,A88,L87]. The least fixpoint of T_P is denoted by lfp(T_P), whereas M_P will stand for the minimal Herbrand model of a theory P.

Definition 2.1 *(Union & intersection)*
Given two theories P and Q:
- **P∪Q** is a theory-valued expression which denotes the theory obtained by putting the clauses of theories P and Q together (juxtaposition);
- **P∩Q** is a theory-valued expression which denotes the theory obtained from P and Q in the following way.

If	$p(t_1,...,t_n) \leftarrow Body_1$	is a clause of P
and	$p(u_1,...,u_n) \leftarrow Body_2.$	is a clause of Q
and	$\exists \vartheta = mgu((t_1,...,t_n),(u_1,...,u_n))$	
then	$p((t_1,...,t_n)\vartheta) \leftarrow (Body_1,Body_2)\vartheta$	is a clause of P∩Q. □

Union and intersection provide two basic theory composition mechanisms. On one hand, union yields a combined theory where the original ones cooperate each other during deduction, in the sense that partial conclusions of either theory may possibly serve as premises for the other theory. On the other hand, intersection yields a combined theory where both the original theories are forced to agree during deduction, in the sense that they must agree on every single partial conclusion (i.e. at each step of the computation).

A logic justification for the former definition stems from the completed version of a logic theory [C78], where union and intersection correspond respectively to disjunction and conjunction [MP88]. More precisely, suppose that predicate p is defined by the formula p↔Γ in the completed version of the theory P, and that the same predicate is defined by the formula p↔Δ in the completed version of the theory Q. Then, predicate P is defined by the formula p↔Γ∨Δ in the completed version of the theory P∪Q, and by the formula p↔Γ∧Δ in the completed version of the theory P∩Q.

The following section shows how to extend the results of [MP88] by introducing a model-theoretic characterization of the two operators.

3 Model-theoretic semantics

In this section the minimal Herbrand model of P∪Q and P∩Q is defined in terms of some classes of models related to P and Q. To do this, we exploit some results of [MP88] relating the immediate consequence operator $T_{P \cup Q}$ and $T_{P \cap Q}$ to the ones associated with T_P and T_Q, which are summarized in the following:

Proposition 3.1 [MP88]
Given two theories P and Q:
i) $T_{P \cup Q} = \lambda I.\ T_P(I) \cup T_Q(I)$
ii) $T_{P \cap Q} = \lambda I.\ T_P(I) \cap T_Q(I)$ □

In the sequel, if P is a logic program and ground(P) is is the set of all ground instances of clauses in P, the notation P'«P stands for "P' is a subset of ground(P)" and facts(P) denote the set of unit clauses of P.

Definition 3.2 *(Model theoretic semantics of $\cup$ and $\cap$)*
Given two theories P and Q,
$M_{P \cup Q} = \cap\ \{M \mid M \mid= P\ \&\ M \mid= Q\} = \min\ \{M \mid M \mid= P\ \&\ M \mid= Q\}$
$M_{P \cap Q} = \cup\ \{M \mid M = M_{P'} = M_{Q'}\ \&\ P'«P\ \&\ Q'«Q\ \&\ facts(P')=facts(Q')\}$ □

The above definition states that the minimal model of the union of two theories P and Q is the minimal interpretation which is a model for both P and Q. On the other hand, the minimal model of the intersection of two theories is the greatest common minimal model of two ground subtheories of P and Q with the same set of unit clauses.

The following propositions 3.3 and 3.5 prove the above claim, namely that $M_{P \cup Q}$ and $M_{P \cap Q}$ are indeed the minimal Herbrand models of $P \cup Q$ and $P \cap Q$ respectively.

Proposition 3.3
Given two theories P and Q:
        $lfp(T_{P \cup Q}) = M_{P \cup Q}$
Proof
By mutual inclusion.
i) $lfp(T_{P \cup Q}) \subseteq M_{P \cup Q}$. In fact, $M_{P \cup Q}$ is a model of both P and Q by definition and hence $T_P(M_{P \cup Q}) \subseteq M_{P \cup Q}$ and $T_Q(M_{P \cup Q}) \subseteq M_{P \cup Q}$.
As a consequence, $T_{P \cup Q}(M_{P \cup Q}) = T_P(M_{P \cup Q}) \cup T_Q(M_{P \cup Q}) \subseteq M_{P \cup Q}$, that is $M_{P \cup Q}$ is a Herbrand model of $P \cup Q$ and hence it contains the minimal Herbrand model, $lfp(T_{P \cup Q})$.
ii) $M_{P \cup Q} \subseteq lfp(T_{P \cup Q})$ since $lfp(T_{P \cup Q})$ is a model of both P and Q. □

Lemma 3.4 is used in the proof of proposition 3.5, and states that if two ground theories share the same minimal model and the same set of facts, then their intersection has the same minimal model.

Lemma 3.4

Given two ground theories P_1 and Q_1 such that $facts(P_1)=facts(Q_1)$ and $M=M_{P_1}=M_{Q_1}$. Then $M_{P_1 \cap Q_1}=M$.

Proof

Observe that $M_{P_1 \cap Q_1} \subseteq M_{P_1}$ since, by [MP88] $M_{P_1 \cap Q_1} \subseteq M_{P_1} \cap M_{Q_1}$.

It suffices to show that:

$$T_{P_1 \cap Q_1}\uparrow\omega \supseteq T_{P_1}\uparrow\omega = T_{Q_1}\uparrow\omega$$

We prove now that

for each $n<\omega$: $A \in T_{P_1}^n(\varnothing)$ & $A \in T_{Q_1}^n(\varnothing)$ *implies* $A \in T_{P_1 \cap Q_1}^n(\varnothing)$

by induction on the stage when A becomes a member of both $T_{P_1}^n(\varnothing)$ and $T_{Q_1}^n(\varnothing)$.

case 1): trivial.

case n)

$A \in T_{P_1}^n(\varnothing)$ & $A \in T_{Q_1}^n(\varnothing)$

$\exists A \leftarrow B$ in P_1 such that: $B \subseteq T_{P_1}^{n-1}(\varnothing)$

$\exists A \leftarrow C$ in Q_1 such that: $C \subseteq T_{Q_1}^{n-1}(\varnothing)$

by the inductive hypothesis: $B \cup C \subseteq T_{P_1 \cap Q_1}^{n-1}(\varnothing)$ □

Proposition 3.5

Given two logic theories P and Q:

$lfp(T_{P \cap Q}) = M_{P \cap Q}$

Proof

By mutual inclusion.

i) $lfp(T_{P \cap Q}) \subseteq M_{P \cap Q}$

We prove by induction that for each $n<\omega$, there exist $P_1 \ll P$ and $Q_1 \ll Q$ such that

$\bigcup_{j \leq n} T_{P \cap Q}\uparrow j = M_{P_1}$ and $\bigcup_{j \leq n} T_{P \cap Q}\uparrow j = M_{Q_1}$, which implies i).

For n=0, the statement is trivial. Let us assume that it holds for each $j<n$. Consider now a ground atom A such that $A \in T_{P \cap Q}\uparrow n$ and $A \notin T_{P \cap Q}\uparrow (n-1)$. Then there exists a clause in $P \cap Q$ of the form $H \leftarrow (Body_1, Body_2)\vartheta$ and a grounding substitution γ such that

$A = H\gamma$ and $T_{P \cap Q}\uparrow (n-1) | = (Body_1, Body_2)\vartheta\gamma$.

By definition of $\cap$, $H_1 \leftarrow Body_1$ is a clause of P, $H_2 \leftarrow Body_2$ is a clause of Q, $\vartheta=mgu(H_1,H_2)$ and $H = H_1\vartheta$. By the inductive hypothesis, we know that there exist $P_1 \ll P$ and $Q_1 \ll Q$ such that $\bigcup_{j<n} T_{P \cap Q}\uparrow j = M_{P_1}$ and $\bigcup_{j<n} T_{P \cap Q}\uparrow j = M_{Q_1}$. Consider the ground theory P'_1 (resp. Q'_1) obtained by adding to P_1 (resp. Q_1) the ground clause $(H_1 \leftarrow Body_1)\vartheta\gamma$ (resp. $(H_2 \leftarrow Body_2)\vartheta\gamma$) for each such atom A. Obviously, $P'_1 \ll P$ and $Q'_1 \ll Q$. We prove that $M=\bigcup_{j \leq n} T_{P \cap Q}\uparrow j = M_{P'_1}=M_{Q'_1}$. First of all, by construction, M is a model of P'_1 and Q'_1. Suppose that it is not minimal, i.e. there exists M' such that $M \supset M'$ and M' is either a model of P'_1 or a model of Q'_1. Let A be a ground atom such that $A \in M$ and $A \notin M'$. Notice that $A \in \bigcup_{j<n} T_{P \cap Q}\uparrow j$ since $\bigcup_{j<n} T_{P \cap Q}\uparrow j$ is the minimal Herbrand model of P_1 and Q_1, and M' includes it. Thus $A \in T_{P \cap Q}(T_{P \cap Q}\uparrow (n-1))$ and $A \notin T_{P \cap Q}\uparrow j$ for each $j<n$. By

construction of P'_1, say, there is a clause $A \leftarrow Body_1 \vartheta \gamma$ in P'_1 such that $M' \models Body_1 \vartheta \gamma$ and $A \notin M'$, contradicting the hypothesis that M' is a model.

ii) $M_{P \cap Q} \subseteq lfp(T_{P \cap Q})$

Notice that — for lemma 3.4 — given two ground theories P_1 and Q_1 such that $M = M_{P_1} = M_{Q_1}$ then $M_{P_1 \cap Q_1} = M$. Let now $A \in M_{P \cap Q}$. Then there exists (by definition of $M_{P \cap Q}$) $P_1 \ll P$ and $Q_1 \ll Q$ such that $M_{P_1} = M_{Q_1} = M'$ and $A \in M'$. By the previous observation, $M' = M_{P_1 \cap Q_1}$. Then ii) holds by observing that $M_{P_1 \cap Q_1} \subseteq M_{P \cap Q}$. $\qquad \square$

4   Meta-interpretive definition

In this section, a meta-interpretive definition of the union and intersection operators described so far is introduced. The standard *solve* (vanilla) meta-interpreter, which simulates the computational model of logic programs, is usually defined as follows [HL89]:

```
solve (true)    ←
solve (G1,G2)  ←  solve (G1), solve (G2)
solve (G)      ←  clause(G←B), solve (B)
```

where the predicate *clause* is used to represent the object program. In moving from a single-theory towards a multi-theory framework, where many logic programs can be possibly combined together to form a more complex program, the *solve* meta-interpreter has to be suitably extended. To this purpose, an extra argument is introduced to specify the program where to prove a certain goal.

```
solve (T , true)      ←                                        (s1)
solve (T , (G1,G2))  ←  solve (T , G1), solve (T , G2)         (s2)
solve (T , G)        ←  clause(T , G←B), solve (T , B)         (s3)
```

Correspondingly, the definition of *clause* is augmented to denote for each object clause also the theory it belongs to. Theory identifiers are assumed to be unique constant symbols. Thus, a clause $H \leftarrow B$ in the theory identified by foo is represented by clause(foo,H←B).

It is worth noting that the theory argument in *solve* and *clause* is not necessarily bound to a constant theory identifier. Actually, it will be any expression built from theory identifiers and the composition operators. The meta-definition of the theory-valued expressions is given using recursion in the *clause* predicate.

clause $((T1 \cup T2), H \leftarrow B)$ $\leftarrow$ clause$(T1, H \leftarrow B)$ (c1)

clause $((T1 \cup T2), H \leftarrow B)$ $\leftarrow$ clause$(T2, H \leftarrow B)$ (c2)

clause $((T1 \cap T2), H \leftarrow (B1,B2))$ $\leftarrow$ clause$(T1, H \leftarrow B1)$, clause$(T2, H \leftarrow B2)$ (c3)

Let us now prove that the meta-interpretive definition of the union and intersection operators is equivalent to the intensional (transformational) definition. Partial evaluation techniques are exploited to prove that the unfolded version of the solve meta-interpreter is exactly the theory generated by the intensional operators. As shown in [HL89], the soundness and completeness of the *solve* meta-interpreter can be proved by making use of partial evaluation techniques.

Proposition 4.1

Let P and Q be two logic theories. Let T denote the program composed of the solve meta-interpreter and of the clauses of P and Q represented by means of the extended *clause* predicate. Let Op denote either $\cup$ or $\cap$.

The theory obtained by partially evaluating T w.r.t. the goal $\leftarrow$ solve$((P$ Op $Q), p(X))$, for any predicate p defined in P or in Q, is P Op Q.

Proof

i) Op=$\cap$.

Consider a predicate p defined both in P and in Q and the meta-goal: solve$((P \cap Q) , p(X))$.

The first step consists of unfolding the goal by means of the atomic case of the solve definition (i.e. the third clause of solve). Unfolding steps are denoted by "$\Rightarrow$".

solve$((P \cap Q) , p(X))$

$\Rightarrow$ *by (s3)*

clause$((P \cap Q), p(X), G)$, solve $((P \cap Q) , G)$

$\Rightarrow$ *by (c3)*

clause$(P, p(X) \leftarrow G_1)$, clause$(Q, p(X) \leftarrow G_2)$, solve $((P \cap Q) , (G_1, G_2))$

$\Rightarrow$ *if $\exists\, p(t) \leftarrow A_1, \ldots, A_m \in P$ such that $\exists\, \gamma = mgu(p(X), p(t))$*

clause$(Q, p(t) \leftarrow G_2)$, solve $((P \cap Q) , ((A_1, \ldots, A_m), G_2))$

$\Rightarrow$ *if $\exists\, p(t') \leftarrow B_1, \ldots, B_n \in Q$ such that $\exists\, \vartheta = mgu(p(t), p(t'))$*

solve $((P \cap Q) , ((A_1, \ldots, A_m)\vartheta, (B_1, \ldots, B_n)\vartheta))$

$\Rightarrow$ *by (s2), applied m+n times*

solve $((P \cap Q) , (A_1)\vartheta), \ldots,$ solve $((P \cap Q) , (B_n)\vartheta)$

Thus, given a predicate p(X) defined both in P and in Q, for each pair of clauses: $p(t) \leftarrow A_1, \ldots, A_m \in P$ and $p(t') \leftarrow B_1, \ldots, B_n \in Q$ such that $\exists\, \vartheta = mgu(p(t_i), p(t_i'))$, the partial evaluation process generates a clause:

solve$((P \cap Q) , p(X)) \leftarrow$ solve $((P \cap Q) , (A_1)\vartheta), \ldots,$ solve $((P \cap Q) , (B_n)\vartheta)$.

It is straighforward to observe that, by iterating the partial evaluation process for all the predicates p defined in P and in Q, we get the theory $(P \cap Q)$, where each clause, $A \leftarrow A_1,...,A_h$ say, is denoted by:

$solve((P \cap Q), A) \leftarrow solve((P \cap Q), A_1),...,solve((P \cap Q), A_h)$.

ii) Op=$\cup$.

Straightforward from i).   □

The former proposition holds more generally in the case that the theory argument in a *solve* goal is any (possibly non-ground) expression over the algebra built on theory identifiers with the composition operators. This is what the following proposition states.

Proposition 4.2

Let T be any (possibly non-ground) theory-valued term, and G be a goal. Assume that the meta-level goal $\leftarrow solve(T, G)$ succeeds with computed answer substitution ϑ, and that T' is a ground theory-valued term such that $T' \leq (T)\vartheta$.

Then the goal $(G)\vartheta$ can be proved in the theory T'.

Proof

The proof is a direct structural induction on T', using Prop. 4.1 for the inductive step.   □

Notice that, in the hypotheses of the above result, if $(T)\vartheta$ is not ground then its variables can be instantiated with arbitrary theory-valued terms, e.g. with the empty theory, as any ground instance of $(T)\vartheta$ is able to prove the goal $(G)\vartheta$. This remark, together with the observation that variables in theory-valued terms may be left only when computing the application of a union operator, allows one to simply drop the variables from the final answer $(T)\vartheta$.

A naive form of hypothetical reasoning is a first, direct application of the former interpretive definition. Suppose one aims at determining a set of hypotheses which, when added to a given knowledge base kb, allows to prove a given goal g. In this simple framework, viewpoints can be encapsulated into separate theories, to be possibly composed by means of the theory-valued operators. This task can be accomplished by a meta-level goal of the kind $\leftarrow solve(kb \cup X, g)$, which will possibly succeed by binding X to an extension of kb where g is provable. For instance, in the oversimplified situation of three theories: $t_1 = \{p \leftarrow q\}$, $t_2 = \{q \leftarrow\}$ and $t_3 = \{r \leftarrow\}$, represented in the *clause* notation as follows :

    clause $(t_1, p \leftarrow q) \leftarrow$
    clause $(t_2, q \leftarrow) \leftarrow$
    clause $(t_3, r \leftarrow) \leftarrow$

it is possible to determine a theory that, when added to t_1, allows one to prove the goal p, by means of the goal $solve(t_1 \cup X, p)$, which succeeds, among other possibilities, with $X = t_2$. More generally, the goal $solve(X, p)$, actually constructs (compositions of) theories where p is provable, and succeeds, among other possibilities, with $X = t_1 \cup t_2$.

5   Difference between theories

So far, we have built a tool-kit for manipulating theories which includes only union and intersection as operators. On the other hand, our ambition is to provide a rich set of combinators, which allow one to build new knowledge intensive programs out of existing ones. In this respect, it is apparent that at least one class of operators is missing: operators for retracting knowledge from a theory [BK82, HL88]. As in the case of union and intersection, our wish is to find *a transformation of the syntax of a theory which realizes a well defined semantic operation*. Again, the basic semantic framework we refer to is the immediate consequence operator.

This is a neat departure from the traditional approach to retraction in logic languages, where textual deletion of clauses is performed (see [HL88]). For instance, given the theory:

p ← q
q ←
p ←

the retraction of p ← yields the theory:

p ← q
q ←

whereas a more semantically sensible operator should return a theory where p is no longer provable.

The first, obvious, consideration is that computability theory establishes that a general operator which yields a theory corresponding to the difference between the minimal models of two given theories is not computable. In this section two operators of limited scope are introduced, together with some applications which point out their use. It is clear, however, that more refined solutions require the ability of handling negation and, in this framework, the ability of handling forms of constructive negation [BaMPT90, C88]. These issues will be discussed at the end of the paper.

5.1   Retraction of predicate definitions

The first operator we wish to define is one which drops all the clauses defining a given predicate p in a theory Q.
The operator will be denoted by Q–p, where Q is a theory and p the name of a predicate. The semantics of this operator in terms of the immediate consequence operator is quite straightforward and, more importantly, it can be given in terms of set-difference between set of literals, thus maintaining the definitional flavor of the union and intersection operators.In the sequel, \ denotes set difference.

Definition 5.1

$T_{Q-p} = \lambda I.\ T_Q\ (I) \setminus \{L \mid L$ unifies with $p(\_)\}$ □

The transformational definition of the operator is also straightforward.

Definition 5.2

Given a theory Q and a predicate symbol p, Q–p is a theory valued expression, which denotes the theory obtained from Q in the following way.

If	$q \leftarrow$ Body	is a clause of Q
and	$q \neq p$	
then	$q \leftarrow$ Body	is a clause of Q–p. □

Despite its simplicity, this operator allows one to design some interesting applications. For example it is possible to use the above defined operator for a formal reconstruction of some hierarchical relationships among chunks of knowledge, represented by logic theories (see [BrMPT90]).

Intuitively speaking, a hierarchical relation between two theories states that the predicate definitions contained in a SUPER theory are inherited by, i.e. become visible to, a SUB theory which is hierarchically linked to the former. Typically, an exception mechanism is adopted, which states that the SUB theory inherits from the SUPER theory everything but what itself re-defines.

In the following, given two theories T_i and T_j, $preds_{ij}$ will stand for the set of predicates which are defined in both theories. Two kinds of theory valued operators are introduced in order to model hierarchical relationships between T_i and T_j.

- **is_a(T_i, T_j):**
  T_i is the SUB theory and T_j is the SUPER theory. The resulting theory contains all the predicate definitions of T_i and inherits the definitions of predicates occurring in T_j only. Notice that the definitions in T_j of predicates in $preds_{ij}$ are replaced by the corresponding definitions in T_i.

- **constraint(T_i, T_j):**
  T_i is the SUB theory and T_j is the SUPER theory. The resulting theory contains the intersection of the definitions of predicates in $preds_{ij}$ and inherits the definitions of predicates occurring in T_j only.

Let us give two simple examples of the use of these operators. Let Person be the following theory:

```
Person = {eats(meat) ←
          eats(fish) ←
          eats(vegetables) ←
          lives_in(x) ← works_in(x)
          age(x) ← current_year(y), born_on(w), x=y-w
          ...
       }
```

The fact that John is a vegetarian working in London can be described by the following is_a link:

```
John =is_a ( {eats(vegetables) ←
              works_in(London)←},
              Person)
```

On the other hand, to describe adult people we can exploit the constraint operator as follows:

Adults = constraint({age(x) ← x≥14, x≤70}, Person)

Let us now formally define the is_a and constraint operators in terms of the basic operators on theories.

Definition 5.3
Given two theories T_1 and T_2,

$$is\_a(T_1, T_2) = T_1 \cup (T_2 - preds_{12}) \qquad \square$$

Notice that the operator – has been used in a generalized way by applying it to a theory and a set of predicate names.

Definition 5.4
Given two theories T_1 and T_2,

$$constraint(T_1, T_2) = (T_1 \cap T_2) \cup (T_2 - preds_{12}) \qquad \square$$

The semantics of the hierarchical operators is directly entailed by the basic theory composition operators exploited in their definition.

5.2   Retraction of theories

The retraction operator, which has been defined in the previous subsection, provides a too coarse granularity for some classes of applications. To be more specific, it works at the whole predicate definition level. This approach, although providing some forms of default reasoning, does not accommodate the handling of exceptions [KS90].

In this subsection we discuss an extension of the above difference operator, which retracts from a theory another theory which has to be a set of facts, i.e. unit clauses. The semantics we expect for such an operator is given by the following definition.

Definition 5.5

Given two theories P and Q, where Q contains only unit clauses
$$T_{P\backslash Q} = \lambda I.\ T_P(I) \backslash \{L \mid \exists\ L' \in Q\colon\ L \text{ unifies with } L'\} \qquad \square$$

Notice that this definition is equivalent to
$$T_{P\backslash Q} = \lambda I.\ T_P(I) \backslash T_Q(I)$$
as T_Q is the constant function $\lambda I.\ I'$, where I' is the set of ground instances of the unit clauses in Q.

The transformational definition of the operator needs a limited form of constructive negation [BaMPT90,C88]. In particular we need a constructive negation procedure for unit clauses, which is performed according to the following definition.

Definition 5.6 *(Negation of a unit clause)*
Let $p(t_1, \dots, t_n)$ be a unit clause. Then its constructive negation, which is denoted by $(p(t_1, \dots, t_n))^-$ is the theory formed by the following clauses:

$$p(x_1, \dots, x_n) \leftarrow x_1 \neq t_1.$$
$$\dots\dots\dots\dots\dots$$
$$p(x_1, \dots, x_n) \leftarrow x_n \neq t_n.$$

where $\neq$ stands for "not equal", and $t_1, \dots, t_n$ is an unrestricted tuple of terms, i.e. one where each variable occurs exactly once. The definition in the case of restricted tuples is omitted here for the sake of simplicity [BaMPT90]. $\qquad \square$

The operational meaning of $\neq$ can be given in different ways. For example Chan uses inequalities as constraints and possibly returns equality formulas as part of the (qualified) answers, by exploiting suitable normalization algorithms. On the other hand, in [BaMPT90] a method is provided for replacing an inequality of the form $x \neq t$ by a disjunction of equalities of the form $x = t$ and inequalities of the form $x \neq y$, where x and y are variables and t is a term which is not a variable.

We are now in the position of defining the constructive negation of a set of unit clauses.

Definition 5.7 *(Negation of a set of unit clauses)*
Let P be a theory of unit clauses only. For each predicate symbol p occurring in P, let

$$p(t_{11},...,t_{1n}) \leftarrow$$
$$\cdots\cdots\cdots\cdots\cdots$$
$$p(t_{m1},...,t_{mn}) \leftarrow$$

be the m clauses defining the n-ary predicate p. Consider the following set, C_p, of clauses:

$$p(x_1,...,x_n) \leftarrow x_{i_1} \neq t_{1i_1},...,x_{i_m} \neq t_{mi_m}.$$

where $\{i_1,...,i_m\}$ is a bag of indexes from $[1,n]$. Then the negation of P, denoted by P^-, is defined as the collection of clauses C_p for each predicate symbol p in P.   □

Put another way, C_p is built by defining clauses, each of which contains m inequalities in its body and each inequality falsifies exactly one of the original clauses. Notice that C_p can be equivalently obtained using the intersection operator in the following way:

$$(p(t_{11},...,t_{1n}))^- \cap ... \cap (p(t_{m1},...,t_{mn}))^-$$

The transformational definition of \ is the following.

Definition 5.8

Let P be a logic theory and Q a theory of unit clauses. Then $P \setminus Q$ is a theory valued expression, denoting the theory obtained in the following way:

$$P \backslash Q = P_1 \cup (P_2 \cap Q^-)$$

where $P = P_1 \cup P_2$ and P_1 (resp. P_2) contains the clauses of the predicates which are not (resp. are) defined in Q.   □

It is now easy to show that Def. 5.8 correctly implements the retraction operator, as defined in Def. 5.5.

Proposition 5.9

$T_{P\backslash Q} = T_{P_1 \cup (P_2 \cap Q^-)}$

Proof

We calculate:

$T_{P_1 \cup (P_2 \cap Q^-)}(I)$

= *by Prop. 3.1*

$T_{P_1}(I) \cup (T_{P_2}(I) \cap T_{Q^-}(I))$

= *by the construction of Q^- and since Q contains only unit clauses,*
  denoting by K the set theoretic complement of $T_Q(I) = ground(Q)$

$T_{P_1}(I) \cup (T_{P_2}(I) \cap K)$

= *by taking into account that $T_{P_1}(I)$ and K are disjoint*

$(T_{P_1}(I) \cup T_{P_2}(I)) \cap K$

= *because $T_{P_1}(I) \cup T_{P_2}(I) = T_P(I)$*

$T_P(I) \cap K$

= $T_P(I) \setminus T_Q(I)$

$=$ *by Def 5.5*
   $T_{P\backslash Q}(I)$

$\square$

It is immediate to see that the retraction operator "−" can be defined using the retraction operator "\" which has been given here.

Proposition 5.10
$Q{-}p \;=\; Q \setminus \{p(X) \leftarrow \}$

$\square$

As in the case of $\cup$ and $\cap$, a meta-interpretive counterpart for \ is provided by suitably extending the *clause* predicate.

Definition 5.11
   clause ((T1\T2), H, B) $\leftarrow$ clause((T1$\cap$T2$^-$), H, B)

where T2$^-$ is the application to T2 of the basic operator for negating a set of unit clauses, as in Def. 5.7.

$\square$

As an example , we show how this retraction operator allows one to build theories with exceptions. Given a theory T, which encompasses the *normal* knowledge about some problem domain, and given some exceptions, the goal is to obtain a new theory T' in which the exceptions are kept into account. For example, if we have the normal theory of birds defined as:

   Bird = { no_of_legs(X,2) $\leftarrow$
          flies(X) $\leftarrow$
          ... }

and a special bird tweety with a single leg which does not fly, we would like to obtain a new theory AbBird which behaves as follows. The evaluation of the goal $\leftarrow$flies(tweety) w.r.t. the theory AbBird finitely fails; the evaluation of the goal $\leftarrow$flies(X) w.r.t. the same theory yields the answer X $\neq$ tweety, while the evaluation of the goal $\leftarrow$no_of_legs(X,Y) yields the answers (X=tweety,Y=1) and (X$\neq$tweety,Y=2).

Let us consider, for the sake of simplicity, a single exception at a time. The exception is represented by a pair of theories which, for the time being, contain a single unit clause. The first one (T_{no}) specifies which pieces of information are not valid any more in the general theory. For example, if the exception concerns the number of legs of tweety, T_{no} is:
   no_of_legs(tweety,_) $\leftarrow$

The second theory (T_{yes}) states which positive knowledge characterizes the exceptional case. Going on with the example:

no_of_legs(tweety,1) ←

Intuitively, a theory which keeps an exception into account is built by first taking away from the original theory all the knowledge concerning the exception and, then, adding the positive knowledge. Formally:

Default(T,T_{no},T_{yes}) = (T \ T_{no}) $\cup$ T_{yes}

Continuing the example, we have

AbBird =   Default(Bird, {no_of_legs(tweety,_)},no_of_legs(tweety,1)}) =
                    {no_of_legs(X,2) ← X ≠ tweety.
                    no_of_legs(tweety,1) ←
                    flies(X) ←
                    ...}

Another motivating example for the form of retraction we have proposed is inspired by [BK82]. Database management can be interpreted as a combination of an object language — to represent and query the database — and of meta-language — to update the database as it changes in time. The assimilation of an input sentence into a current database (resulting in a new database) is defined in [BK82] by a logic program including the clause:

assimilate(Currdb, Input, Newdb) ←   Info ∈ Currdb,
                                          Interdb = (Currdb − Info),
                                          demo(Interdb + Input, Info),
                                          assimilate(Interdb, Input, Newdb).

It is trivial to observe that the "+" operator corresponds to our union operator, while the "−" operator is our retraction operator "\", provided that the set of clauses to be retracted, Info, is extensional, i.e. a set of atoms. Intuitively, this type of knowledge assimilation maintains in the program as much intensional knowledge as possible, possibly discarding some extensional definitions which are implied by the rest of the knowledge base. It is worth noting that the knowledge assimilation interpretation of [BK82] can be declaratively modelled by our operators, although in a restricted form (namely the Info theory has to be extensional). Furthermore, the "\" operator does something more with respect to syntactical retraction, in the sense that constraints are added to Interdb such that Info cannot be derived anymore.

In concluding this section we remark that by exploiting the technique of intensional negation defined in [BaMPT90] it is possible to extend the retraction operator "\" to theories containing also clauses, provided that they do not contain local variables, i.e. variables which occur in the body of a clause but not in its head. The expressive power of the operator is obviously increased, although its impact on application is still to be investigated and its formalization makes the whole framework more cumbersome.

6 Conclusions

A repertoire of basic composition operators for logic theories has been introduced, along with their characterizations from different perspectives. Simple examples taken typically from non monotonic reasoning problems have been given to illustrate the use of such operators.

Further research is ongoing towards extending the framework in two main directions. On one hand, moving from pure Horn theories to logic theories with negation is needed, in order to extend the expressiveness and hence the scope for applicability. Although most of the features of the framework extend smoothly to the more general setting, e.g. the transformational and the meta-interpretive characterizations, the model-theoretic one needs further investigation to be adapted. [MPRT90] deals with some of this problems in a restricted class of programs with negation.

On the other hand, the kernel of the basic composition operators presented in this paper should be properly augmented with other (preferably derived) operators, possibly tailored to specific application domains. Examples are modular composition operators which deal with interfaces and in-out specifications, in the spirit of logic-programming-in-the-large, as well as specialized union operators which capture more directly non trivial reasoning mechanisms, e.g. hypothetical and abductive reasoning.

Acknowledgments

This work has been partially supported by ESPRIT Project 3012 "Compulog". We would like to thank Bob Kowalski and John Lloyd for the pertinent helpful suggestions.

References

[A88]
Apt, K.R., "Introduction to Logic Programming", Report CS-R8826 Center for Mathematics and Computer Science, Amsterdam, 1988. To appear in: J. van Leeuwen, (ed.), *Handbook of Theoretical Computer Science*, North Holland.

[BaMPT90]
Barbuti, R., Mancarella, P., Pedreschi, D. and Turini, F. "A Transformational Approach to Negation in Logic Programming", *Journal of Logic Programming* **8** pp. 201-228, (1990).

[BK85]
Bowen, K.A., Kowalski, R.A. "Amalgamating Language and Metalanguage in Logic Programming". In *Logic Programming* (K.L. Clark and S.-A. Tarnlund eds.) pp. 153-172 (1985).

[BrMPT90]
Brogi, A., Mancarella, P., Pedreschi, D., Turini, F. "Hierarchies through Basic Meta-Level Operators". In *Proceedings META 90 - Workshop on Meta-Programming in Logic*, pp. 381-396, Leuven, 1990.

[BG77]
Burstall, R.M. and Goguen, J.A. "Putting theories together to make specifications". In *Proceedings 5th IJCAI*, Cambridge, Massachussets, 1990.

[C78]
Clark, K.L. "Negation as Failure". In: H. Gallaire and J. Minker (eds.), *Logic and Data Bases*, Plenum Press, New York., pp. 293-322 (1978).

[C88]
Chan, D. "Constructive Negation Based on the Completed Database". *Proceedings 5th ICLP* (R.A. Kowalski and K. Bowen, eds.), The MIT Press, Cambridge, Mass., pp. 111-125 (1988).

[vEK76]
van Emden,M.H., Kowalski,R.A. "The Semantics of Predicate Logic as a Programming Language". *Journal of the ACM* 23, 733–742, 1976.

[GS89]
Gaifman, H. and Shapiro, E. "Fully Abstract Compositional Semantics for Logic Programs", Proceedings POPL89, ACM, pp. 134-142 (1989).

[HL88]
Hill, P.M. and Lloyd, J.W. "Meta-Programming for Dynamic Knowledge Bases". TR-CS-88-18, Dpt. of Computer Science, University of Bristol (1988).

[HL89]
Hill, P.M. and Lloyd, J.W. "Analysis of Meta-programs". In *H.D. Abramson and M.H. Rogers, editors, Meta-Programming in Logic Programming*, pages 23-52, MIT Press (1989).

[K79]
Kowalski, R.A. *Logic for Problem Solving*. North-Holland Elsevier (1979).

[KS90]
Kowalski, R.A. and Sadri, F., "Logic Programs with Exceptions", in *Proceedings 5th ICLP* (D.H.D. Warren and P. Szeredi, eds.), The MIT Press, Cambridge, Mass., pp. 598-613 (1990).

[L87]
Lloyd, J.W., *Foundations of Logic Programming*, Springer Symbolic Computation Series, Berlin, (1987).

[LM84]
Lassez, J.-L. and Maher, M. "Closures and fairness in the semantics of logic programming". In *Theoretical Computer Science*, 29: 167-184 (1984).

[MP88]
Mancarella, P. and Pedreschi, D. "An Algebra of Logic Programs". in *Proc. of Fifth International Conference, Symposium of Logic Programming*, Seattle 1006-1023 (1988).

[MPRT90]
Mancarella, P., Pedreschi, D., Rondinelli, M. and Tagliatti, M. "Algebraic Properties of a Class of Logic Programs". In *Proc. of Second NACLP*, Austin (1990). To appear.

The Synthesis of Logic Programs from Inductive Proofs *

Alan Bundy   Alan Smaill   Geraint Wiggins

September 12, 1990

Abstract

We describe a technique for synthesising logic (Prolog) programs from non-executable specifications. This technique is adapted from one for synthesising functional programs as total functions. Logic programs, on the other hand, define predicates. They can be run in different input modes, they sometimes produce multiple outputs and sometimes none. They may not terminate. The key idea of the adaptation is that a predicate is a total function in the all-ground mode, *i.e.* when all its arguments are inputs ($pred(+, \ldots, +)$ in Prolog notation). The program is synthesised as a function in this mode and then run in other modes. To make the technique work it is necessary to synthesise pure logic programs, without the closed world assumption, and then compile these into Prolog programs. The technique has been tested on the OYSTER (functional) program development system.

1   Introduction

The ideal aspired to in logic programming is allow computer users to describe their problem in the language of predicate logic. A clever interpreter will then run their logical description as a computer program and this will solve the user's original problem. Hence computer users will be freed from the necessity of thinking of their problems in procedural terms.

Current logic programming languages do not realise this ideal, [Bundy 88a]. The logical specification of a computer program may fail to be executable either efficiently or at all. For instance, it may not be in clausal form. The Lloyd-Topor translation process ([Lloyd 87], p113) will put it in clausal form, but the presence of negations in the clause body may cause it to flounder when executed using negation as failure. It may contain non-constructor functions, which need to be turned into predicates before they can be executed. The algorithm produced by a direct execution of the specification may be hopelessly inefficient.

---

*The research reported in this paper was supported by Esprit BRA grant 3012, and an SERC Senior Fellowship to the first author. We are grateful for feedback from Frank van Harmelen, David Basin and an anonymous referee on earlier drafts. Seán Matthews helped us defeat TEX.

One answer to these difficulties is to transform the original specification into an equivalent logical formula which can be executed efficiently. A variety of such transformation techniques have been proposed, *e.g.* [Hogger 81, Bruynooghe *et al* 89]. In this paper we discuss how to adapt a technique for synthesising functional programs from logical specifications to the synthesis of logic programs. This technique is based on the 'proofs as programs' paradigm, and is implemented, for instance, in the Nuprl program development system, [Constable *et al* 86]. It is useful to consider the 'proofs as programs' technique for two main reasons.

- It is a powerful technique, which is useful for synthesis, transformation and verification. Adapting it to logic programming might well reveal extensions to current techniques. For instance, since it is based on higher order, typed logics, it will suggest how to adapt logic program transformation to such logics. It relates proof structure to program structure and, hence, program efficiency, thus providing a logical account of computational complexity, which can be used to guide program transformation.

- It is has a well developed theoretical foundation. Relating the 'proofs as programs' technique to existing logic programming transformation techniques might help us to understand them better. In particular, it relates recursive programs and inductive proofs in an intimate way. This may help us understand 'loop spotting' techniques, such as those in [Bruynooghe *et al* 89], as a form of inductive proof.

The adaption of the 'proofs as programs' technique is not straightforward. It is only able to synthesise programs that are *total functions*, that is programs that are defined for all inputs of the right type and are guaranteed to terminate and to return precisely one output. On the other hand, the declarative meaning of a logic program procedure is a *predicate*. These predicates may be called as procedures in a variety of different input modes. For some combinations of input they fail and return no output, for others they return more than one output on backtracking. They are not guaranteed to terminate. The challenge is to adapt the 'proofs as programs' technique to synthesise these completely different kinds of mathematical objects.

2   The Proofs as Programs Technique

We begin by describing the proofs as programs technique. We have implemented this in the OYSTER system, [Horn 88], which is a re-implementation of Nuprl in Prolog. OYSTER and Nuprl are interactive theorem provers for a logic based on Martin-Löf Intuitionistic Type Theory, [Martin-Löf 79], a higher-order, richly typed logic. Using OYSTER, programs are synthesised from their specifications by proving a *specification theorem* of the form:

$$\forall Inputs, \exists Output. \ spec(Inputs, Output)$$

where *spec(Inputs, Outputs)* is a relationship between the inputs and the output of the desired program. This theorem is proved constructively and the resulting proof is analysed to extract the implicit algorithm it defines for calculating the required output given any combination of inputs. A constructive proof is required to avoid the possibility of a pure existence proof in which the existence of an output is proved without any implicit algorithm being defined. OYSTER provides an interactive proof editor, which allows the user to guide the process of proof construction.

For instance, the specification of set union could be written as[1]:

$$\forall A{:}\mathcal{U}, \ \exists ASets{:}\mathcal{U}, \ \forall S_1{:}ASets, \ \forall S_2{:}ASets, \ \exists S_3{:}ASets,$$
$$\forall El{:}A(El \in S_3 \leftrightarrow El \in S_1 \vee El \in S_2)$$

where $X{:}\tau$ means X is an object of type τ, A is some type of objects, $ASets$ is the type of finite sets of such objects and $\mathcal{U}$ is the type of all simple types[2].

OYSTER's Type Theory is an especially suitable logic for the task of program synthesis because it not only provides a constructive logic, as required, but it greatly simplifies the task of extracting the program from the proof. Every rule of inference of the logic has an associated rule of program construction, so that the program is constructed as the proof progresses. Thus there is a duality between proof steps and program steps, for instance applications of mathematical induction in the proof create recursion in the program. The synthesised program is also in OYSTER's logic, and can, therefore, be interpreted as a higher-order, typed, functional program. This program is called the *extract term*.

Because the logic is typed, the type of each variable, in the example specification of union above, has to be declared in the variable's quantification. OYSTER uses these type declarations to do synthesis time type checking, rather than run time or compile time type checking. The function synthesised is only guaranteed to meet the specification when applied to sets of objects of type A. We will indicate this by putting a subscript on the function name, *i.e.* $\cup_A$.

3   An Example of Program Synthesis

The process of program synthesis via theorem proving is illustrated by the following example, using the specification of set union given above, (2).

After some elimination of quantifiers the state of the proof might be:

$$a{:}\mathcal{U}$$
$$s_1{:}sets(a)$$
$$s_2{:}sets(a)$$
$$\vdash_\Theta \ \exists S_3{:}sets(a), \ \forall El{:}a(El \in S_3 \leftrightarrow El \in s_1 \vee El \in s_2)$$

where the extract term constructed so far is $\lambda a, \lambda s_1, \lambda s_2.\Theta$, where Θ is the extract term to be constructed from the remainder of the proof. This suggests a program definition of:

$$S_1 \cup_A S_2 \ = \ \Theta$$

---

[1] In order to make these examples intelligible to an audience unfamiliar with intuitionistic type theory we have used standard logical notation rather than that used in OYSTER. We follow the Prolog convention that identifiers starting with capital letters are variables.

[2] $\mathcal{U}$ is not itself a simple type. If it were we would fall foul of Russell's paradox.

Suppose we now decide to apply induction on s_1. This will produce subgoals corresponding to the base case and the step case of the induction.

$$a:\mathcal{U}$$
$$s_2:sets(a)$$
$$\vdash_\Phi \quad \exists S_3:sets(a), \forall El:a(El \in S_3 \leftrightarrow El \in \emptyset \vee El \in s_2)$$

$$a:\mathcal{U}$$
$$el':a$$
$$s_1:sets(a)$$
$$s_2:sets(a)$$
$$\exists S_3:sets(a), \forall El:a(El \in S_3 \leftrightarrow El \in s_1 \vee El \in s_2)$$
$$\vdash_\Psi \quad \exists S_3:sets(a), \forall El:a(El\ instS_3 \leftrightarrow El \in el' \circ s_1 \vee El \in s_2)$$

where $el' \circ s_1$ is formed by adding a new member el' to the set s_1. $el' \circ s_1$ is a set if el' is not already a member of s_1. The new state of the program definition is:

$$\emptyset \cup_A S_2 = \Phi$$
$$(El' \circ S_1) \cup_A S_2 = \Psi$$

where Φ and Ψ are the extract terms of the base and step cases of the proof, respectively.

This is enough of the proof to give the flavour of the synthesis technique. The proof of the base and step cases will now proceed and will instantiate Φ and Ψ . The final program might be:

$$\emptyset \cup_A S_2 = S_2$$
$$El' \in S_2 \rightarrow (El' \circ S_1) \cup_A S_2 = S_1 \cup_A S_2$$
$$\neg El' \in S_2 \rightarrow (El' \circ S_1) \cup_A S_2 = El' \circ (S_1 \cup_A S_2)$$

4 Synthesising Logic Programs

How can this technique be adapted to the synthesis of logic programs, *e.g.* programs in Prolog?

Firstly, Prolog is neither higher-order nor typed, so we need to prevent the occurrence of these features in the synthesised programs. This is easily achieved by using a first-order logic in the place of OYSTER's current logic, or by restricting OYSTER's logic to its first order part. We still require this logic to be constructive and to associate program construction rules with each rule of inference. It will also be convenient to use a typed logic during synthesis, and then drop any reference to types in the final program. Otherwise, in order to ensure that a Prolog procedure is defined for all its arguments, we will have to provide clauses to deal with arguments that lie outside the intended types, *e.g.* clauses for *append*/3 for non-list arguments.

Secondly, OYSTER will produce total functions, whereas we require partial, multi-valued and, sometimes, non-terminating predicates. The 'proofs as programs' technique has been extended to synthesise partial and non-terminating functions, *e.g.* by restricting the domain of the function to sub-domains where it is both defined and terminating. The technique has not been extended to multi-valued functions. It is this problem we address in this paper.

5 Compiling Functions

An obvious solution is to synthesise a first-order, total function, then compile it into Prolog. The drawback of this solution is that it produces a Prolog program that is underdefined for some of its arguments and hence in some modes. Suppose we synthesise a function $foo{:}\tau_1 \times \ldots \times \tau_n \mapsto \tau$. This compiles into a Prolog predicate, $foo/n+1$, of type $\tau_1 \times \ldots \times \tau_n \times \tau$. This definition will not necessarily be exhaustive on its last argument, so it may not be defined for modes other than $foo(?,\ldots,?,-)$.

Consider, for instance, a function $double{:}nat \mapsto nat$, defined as:

$$double(0) = 0$$
$$double(s(M)) = s(s(double(M)))$$

This will become the Prolog procedure:

$$double(0,0).$$
$$double(s(M),s(s(N))) \ :- \ double(M,N)$$

This will be fine in mode $double(?,-)$, but not in mode $double(?,+)$. Consider, for instance, the call $double(M,3)^3$. This will fail. Maybe this is what was intended, but one cannot be sure of this since the full range of inputs for the second argument was not considered during the synthesis of the procedure. If $double$ were originally defined as a predicate then all the cases would have to have been considered, maybe coming up with a definition like:

$$
\begin{aligned}
double(0,0) &\leftrightarrow true \\
double(s(M),0) &\leftrightarrow false \\
double(0,s(0)) &\leftrightarrow true \\
double(s(M),s(0)) &\leftrightarrow false \\
double(0,s(s(N))) &\leftrightarrow false \\
double(s(M),s(s(N))) &\leftrightarrow double(M,N)
\end{aligned}
\tag{1}
$$

which will find the integer half in mode $double(-,+)$, *e.g.* the call $double(M,3)$ will instantiate M to 1. Note case (1). By giving this the body *true* we ensure a non-*false* value in mode $double(-,+)$, even when the second argument is an odd number. If we want it to fail in such cases we should give (1) the body *false*. This definition compiles to the Prolog procedure:

$$double(0,0).$$
$$double(0,s(0)).$$
$$double(s(M),s(s(N))) \ :- \ double(M,N)$$

---

[3] Where 3 is shorthand for $s(s(s(0)))$.

6 Predicates as Functions

We explore an alternative solution to this problem of synthesising multi-mode programs. The proofs as programs technique is adapted to synthesise predicates instead of functions, so that the resulting Prolog procedure is defined for all its arguments, and hence all input modes. The key idea of the solution is that in the all-ground mode input mode *i.e. pred*$(+, \ldots, +)$) logic programs *are* total functions. Hence, OYSTER and similar systems can be used directly to synthesise logic programs in this mode. If they are called in another mode then they will not be total functions, but this will not matter.

Suppose foo/n is a Prolog procedure. We will consider foo as an n-ary predicate of first-order typed logic. Let the type of the ith argument of foo be τ_i. foo can also be regarded as a function of type $\tau_1 \times \ldots \times \tau_n \mapsto boole$. Observe that, in the all-ground mode, $foo(+, \ldots, +)$, foo/n is a function. If all its arguments are ground then it must take either the value *true* or *false*. It cannot take both values and it cannot take neither. If foo is also terminating then it will be a *total* function.

For example, consider the procedure $del/3$, which deletes one occurrence of a particular element from a list. Its standard Prolog definition is:

$$del(X, [X|Tl], Tl). \tag{2}$$
$$del(X, [Hd|Tl], [Hd|L']) \; :- \; del(X, Tl, L'). \tag{3}$$

with the intended mode $del(+, +, -)$. This is an archetypal example of a partial and multi-valued procedure. The call $del(a, [b, c], L)$ fails, so no value is found for L. The call $del(a, [a, b, a], L)$ succeeds twice, first with the value $[b, a]$ for L and second with the value $[a, b]$.

However, in mode $del(+, +, +)$, $del/3$ *is* a function. For instance, $del(a, [b, c], [b, c])$ has value *false*; $del(a, [a, b, a], [b, a])$ has value *true* and $del(a, [a, b, a], [a, b])$ has value *true*.

7 Pure Logic Programs

This observation is obscured in Prolog because of the heavy use of the closed world assumption. There are no explicit truth values. Falsity is equated with failure; truth with success. In order to use our program synthesis techniques we must use a slightly different notation for logic program procedures in which truth and falsity are explicit. We will adapt and extend the notation first introduced in [Bundy 88b]. Procedures identified with predicates and are defined by a set of formulae, which we will call *tracts*[4]. There is precisely one tract for each combination of arguments to the procedure. A *pure logic program* is a set of predicates each of which is defined by a set of tracts.

A *tract* has the form:

---

[4]These are what we called *cases* in [Bundy 88b]. Unfortunately, this word is already in use to describe the parts of proofs, *e.g.* step case. We cannot use *clauses* because this word is reserved for describing Prolog programs, which we will want to distinguish from tracts. The Student's English Dictionary describes a 'tract' as a "a short dissertation in which some particular subject is treated", and this seems fairly close to the meaning we intend.

$$Condition \rightarrow (Head \leftrightarrow Body)$$

where *Condition* may be *true*, in which case it is omitted. *Head* is of the form $Pred(Arg_1, \ldots, Arg_n)$, where *Pred* is an n-ary predicate. Each argument of *Pred* is either a recursive argument or a parameter argument. If i is a parameter argument then Arg_i must be a variable. If i is a recursive argument then Arg_i must be a constructor term. A constructor term is either a variable or a constructor function applied to constructor terms. All the variables in *Head* are distinct. *Condition* and *Body* can be any first-order formulae that do not contain non-constructor functions.

The tracts defining a predicate are mutually exclusive and exhaustive, *i.e.* they give precisely one value for each combination of arguments. This is best illustrated by an example. Consider the predicate $del{:}\tau \times list(\tau) \times list(\tau) \mapsto boole$. Its tracts might be:

$$del(X, [], L) \leftrightarrow false \tag{4}$$

$$del(X, [Hd|Tl], L) \leftrightarrow (X = Hd \wedge Tl = L) \vee \tag{5}$$
$$(\exists L'{:}list(\tau) \; L = [Hd|L'] \wedge del(X, Tl, L'))$$

These tracts are exhaustive since the second argument of *del* is of type $list(\tau)$. *del/3* will take the value *true* whenever the right hand side of tract (5) evaluates to *true*, *e.g.* for $del(a, [a, b, c], [b, c])$.

Compare these tracts with the Prolog clauses (2) and (3), above. Case (4) defines *del* when the recursive argument is an empty list. There is no Prolog clause corresponding to this. Case (5) corresponds to clauses (2) and (3): each disjunct to the body of one clause. Note that *Tl* and *L* are *not* identified in the head of the tract (5), as they are in clause (2), rather they are set equal in the corresponding disjunct in the body.

These tracts define a total function when called in mode $del(+, +, +)$. What happens for other modes, *e.g.* $del(+, +, -)$? There is bound to be at least one tract matching each calling pattern, but there may be more than one. This means that *del* will always return a result, but it may return more than one. Sometimes this result will be *false* and sometimes *true*. In addition, it will instantiate some of the variables in the arguments either partially or totally. In those cases where *del* returns *true*, we can regard these instantiations as the answers sought. This will give us the partiality and multi-valuedness that we seek. In general, we will call the truth values returned by a predicate call the *results*. When the result is *true* we will call the instantiations of any unbound variables the *outputs*.

Predicates may return the same result by several different computation routes. If there are an infinite number of computation routes then the predicate is non-terminating. If only a finite number of these routes return the result *false* and an infinite number return the result *true*, then we will call the non-termination *benign*. This terminology reflects the observation that this kind of non-termination is often desired as a way of obtaining an infinite set of outputs. If all the infinite number of routes return the result *false* then we will call the non-termination *malignant*, since this kind of non-termination is rarely desired. Otherwise, there will be some *true* routes and an infinite number of *false* ones. We will call this kind of non-termination *pre-cancerous*, since it is possible[5]

---

[5] With a clever interpreter.

to obtain some outputs before the call turns malignant. If there are an infinite number of *true* routes then it is possible to put off the malignancy indefinitely.

For instance, consider the predicate *is_nat*/1, which in mode *is_nat*(+) tests whether its argument is a natural number.

$$is\_nat(0) \leftrightarrow true$$
$$is\_nat(s(N)) \leftrightarrow is\_nat(N)$$

In mode *is_nat*(+) this predicate is total, *i.e.* single valued and terminating. In mode *is_nat*(−) it is benignly non-terminating, producing the infinite set of outputs: $0, s(0), s(s(0)), \ldots$.

Now consider the predicates *is_list*/1 and *both*/1 defined by:

$$is\_list([]) \leftrightarrow true$$
$$is\_list([H|T]) \leftrightarrow is\_list(T)$$

$$both(X) \leftrightarrow is\_nat(X) \wedge is\_list(X)$$

both/1 terminates in mode *both*(+) but is malignantly non-terminating in mode *both*(−)

8 Compiling Pure Logic Programs

§11 defines a compiler for transforming pure logic programs into Prolog programs. Cases whose body is *false* are omitted. ↔s are turned into : −s. Existential quantifiers in the body are dropped. Conditions are put at the front of bodies and tracts are put into clausal form. Body literals of the form $Var = Term$ are omitted and all occurrences of Var replaced by $Term$. Applying this compiler to the pure logic definition of *del* above produces the Prolog program given in clauses (2) and (3) above, as required.

This procedure is guaranteed to be total in the mode $del(+, +, +)$. However, it would be little use if it could only be called in that mode. How will it behave in other modes? In general, if an argument marked + is used in mode - then the procedure may be over-defined, *i.e.* return multiple results. This may cause it to be multi-valued and/or non-terminating.

The efficiency of a Prolog procedure is mode dependent. In particular, the efficiency may be dependent on the order of the literals in each clause. This problem may be best dealt with during the compilation into Prolog, rather than during synthesis. This is because order is irrelevant for pure logic procedures. It only becomes significant for the literals and clauses of Prolog procedures. The intended mode of use should be an input to the compilation phase (see §11) and influence its outcome.

9 An Example of Predicate Synthesis

We can adapt the proofs as programs technique to the synthesis of pure logic procedures as follows. We prove theorems of the form:

$$\forall X_1 : \tau_1, \ldots, X_n : \tau_n, \exists B : boole.\ spec(X_1, \ldots, X_n) \leftrightarrow B \tag{6}$$

where $spec(X_1, \ldots, X_n)$ specifies an n-ary predicate, foo, in a constructive, first order logic. From a proof of this theorem we can extract a definition of foo/n as a pure logic program.

For instance, to specify the $del/3$ program we might prove the theorem:

$$\forall \tau{:}\mathcal{U}, \forall X{:}\tau, \forall K, L{:}list(\tau), \exists B{:}boole.$$
$$(\exists L_1, L_2{:}list(\tau). \qquad L = L_1 <> L_2 \ \wedge \ K = L_1 <> [X|L_2]) \ \leftrightarrow \ B$$

where $<>$ is the infix list append function.

After eliminating the initial quantifiers and applying list induction to K, this theorem reduces to the base and step cases:

$$\tau{:}\mathcal{U}, x{:}\tau, l{:}list(\tau)$$
$$\vdash_\Phi \ \exists B{:}boole. \ (\exists L_1, L_2{:}list(\tau). l = L_1 <> L_2 \wedge [] = L_1 <> [x|L_2]) \leftrightarrow B$$

$$\tau{:}\mathcal{U}, x{:}\tau, l{:}list(\tau)$$
$$hd{:}\tau, tl{:}list(\tau)$$
$$\forall L{:}list(\tau), \exists B{:}boole.$$
$$\qquad (\exists L_1, L_2{:}list(\tau). \ L = L_1 <> L_2 \ \wedge \ tl = L_1 <> [x|L_2]) \ \leftrightarrow \ B$$
$$\vdash_\Psi \ \exists B{:}boole.$$
$$\qquad (\exists L_1, L_2{:}list(\tau). \ l = L_1 <> L_2 \ \wedge \ [hd|tl] = L_1 <> [x|L_2]) \ \leftrightarrow \ B$$

This proof suggests the following partial definition of $del/3$.

$$del(X, [], L) \ \leftrightarrow \ \Phi$$
$$del(X, [Hd|Tl], L) \ \leftrightarrow \ \Psi$$

and Φ and Ψ are the extract terms of the base and step cases, respectively.

Since for no L_1 and L_2 is it the case that $[] = L_1 <> [x|L_2]$ then the base case reduces to $\exists B : boole.false \leftrightarrow B$. This is proved by instantiating B to $false$, which suggests an instantiation of the extract term, Φ, to $false$. This completes the base case.

The step case requires a proof by cases using:

$$L_1 = [] \ \vee \ \exists Hd_1{:}\tau, Tl_1{:}list(\tau).L_1 = [Hd_1|Tl_1]$$

In the first case the induction conclusion reduces to:

$$\exists B{:}boole. \ (\exists L_2{:}list(\tau). \ l = L_2 \ \wedge \ [hd|tl] = [x|L_2]) \ \leftrightarrow \ B$$

using the rewrite rule $[] <> L \Rightarrow L$. Note that this and the other rewrite rules used in the proof are based on equalities or equivalences and hence valid in both directions. This is necessary for soundness since they are applied under $\leftrightarrow$.

Hence, by instantiating L_2 to l, and applying the substitution axiom to the induction conclusion reduces it to:

$$\exists B{:}boole. \ (hd = x \ \wedge \ tl = l) \ \leftrightarrow \ B$$

This is proved by splitting it into four sub-cases using: $hd = x \lor hd \neq x$ and $tl = l \lor tl \neq l$. In three of these sub-cases $hd = x \land tl = l$ reduces to $false$ and the other sub-case it reduces to $true$. B is instantiated accordingly, to complete this first case. This suggests $hd = x \land tl = l$ as the extract term of this first case.

In the second case the induction conclusion reduces to

$$\exists B{:}boole.$$
$$(\exists Hd_1{:}\tau, Tl_1, L_2{:}list(\tau).$$
$$l = [Hd_1|Tl_1 <> L_2] \quad \land \quad [hd|tl] = [Hd_1|Tl_1 <> [x|L_2]])$$
$$\leftrightarrow \quad B$$

We split this into two sub-cases using:

$$l = [] \quad \lor \quad \exists H'{:}\tau, L'{:}list(\tau).l = [H'|L']$$

Since $[] = [Hd_1|Tl_1 <> L_2]$ is $false$, the first sub-case is readily proved with B instantiated to $false$. The second sub-case reduces to:

$$\exists B{:}boole. \ (\exists Tl_1, L_2{:}list(\tau). \ L' = Tl_1 <> L_2 \ \land \ hd = Hd_1 \ \land \ tl = Tl_1 <> [x|L_2]) \ \leftrightarrow \ B$$

by applying the substitution axiom. We then apply a further case split using: $hd = Hd_1 \lor hd \neq Hd_1$. The second sub-sub-case is readily proved with B instantiated to $false$. The first sub-sub-case reduces to:

$$\exists B{:}boole. \ (\exists Tl_1, L_2{:}list(\tau). \ L' = Tl_1 <> L_2 \ \land \ tl = Tl_1 <> [x|L_2]) \ \leftrightarrow \ B$$

This matches the induction hypothesis by instantiating L to L' and renaming the bound variable L_1 to Tl_1. We can then use the induction hypothesis to prove the induction conclusion and complete this sub-sub-case. The use of the induction hypothesis suggests the recursive call $del(x, tl, L')$ as the extract term of this sub-sub-case and $\exists L'{:}list(\tau). \ l = [hd|L'] \land del(x, tl, L')$ as the extract term of the whole second case.

This completes the whole step case, and suggests for it the extract term:

$$(hd = x \land l = tl) \quad \lor \quad (\exists L'{:}list(\tau). \ l = [hd|L'] \land del(x, tl, L'))$$

Hence the final pure logic program suggested is:

$$del(X, [], L) \ \leftrightarrow \ false$$
$$del(X, [Hd|Tl], L) \ \leftrightarrow \ (X = Hd \land Tl = L) \ \lor$$
$$(\exists L'{:}list(\tau). \ L = [Hd|L'] \land del(X, Tl, L'))$$

as required.

10   Implementation in O$^\mathsf{Y}$STER

We have tested this logic program synthesis process in the O$^\mathsf{Y}$STER system. This is not an ideal vehicle for two reasons.

- The programs extracted are functions in the type theory and require transformation into pure logic programs. In general, it is not possible to make this transformation, because ...

- ... OYSTER's logic is higher order. Thus it is possible to extract non-first-order programs that cannot be transformed into pure logic programs — even from first-order specifications.

However, the test was easily performed, since it required no modifications to OYSTER, and provided supportive evidence for the proposals advanced in this paper. Proofs were obtained by interactive use of OYSTER, during which higher order features in either the specification or extract terms were avoided. The extract term was translated into a pure logic program by hand.

In OYSTER's type theory, propositions are identified with the types of their proofs, so a proposition is true if and only if it is inhabited when considered as a type. This suggests a definition of *boole* as $\mathcal{U}$, the type of simple types. However, this permits specification theorems of the form:

$$\forall X_1:\tau_1, \ldots, X_n:\tau_n, \exists B:boole.\ spec(X_1, \ldots, X_n) \leftrightarrow B$$

to be proved in a trivial way by instantiating B to $spec(X_1, \ldots, X_n)$. This synthesies a predicate foo/n whose definition is equal to the original specification, *i.e.*

$$foo(X_1, \ldots, X_n) \quad \leftrightarrow \quad spec(X_1, \ldots, X_n)$$

which is not what we want.

Instead, we defined *boole* as a type containing precisely two members: *true* and *false*, where *false* is the empty type *void* and *true* is some simple, inhabited type. Note that equality on this type is decidable, so that the law of excluded middle holds for this type, *i.e.*

$$\forall B:boole.\quad B = true \quad \lor \quad B = false$$

Since OYSTER's logic is constructive, this law does not hold in general, in particular, it does not hold for the type $\mathcal{U}$.

This prevents the instantiation of B by $spec(X_1, \ldots, X_n)$, unless $spec(X_1, \ldots, X_n)$ has been reduced to a member of *boole*. In fact, we can only prove the specification theorem by exhibiting a mapping from a proof of B to a proof of $spec(X_1, \ldots, X_n)$, and vice versa. The necessary and sufficient condition for the existence of these mappings is that $spec(X_1, \ldots, X_n)$ be decidable. The role of this is to prove $spec(X_1, \ldots, X_n)$ by exhibiting a decision procedure, namely the synthesised program.

Note that if a formulae is decidable then the law of excluded middle holds for it and we can use this law to make a case split. We used this facility extensively in the proof in §9, *e.g.* $hd = x \lor hd \neq x$ was used in case 1 of the step case. It was not possible to split on $spec(X_1, \ldots, X_n) \lor \neg spec(X_1, \ldots, X_n)$ until it had been shown decidable. Our synthesis proof proceeded by reducing $spec(X_1, \ldots, X_n)$ to a collection of decidable subgoals, using case splits to instantaite B and complete the proof of each subgoals, and then reconstituting their decision predicates into the desired logic program.

OYSTER has been used in this way to prove the example theorem in the last section. The resulting extract term is a function in the type theory. In this case the extract term had a clear correspondence to the desired pure logic program for $del/3$. Unfortunately, it is not possible, in general, to translate such extract terms into pure logic programs. To avoid this problem we are working on a version of OYSTER based on a constructive, first order logic in which the extract terms are pure logic programs.

11 A Prolog Compiler

Once we have completely synthesised a pure logic procedure then we need to translate it into Prolog clauses. This can be done by the following compiler, defined by a series of rewritings.

This compiler may translate two logically equivalent pure logic procedures into two logically equivalent[6], but procedurely non-equivalent, Prolog procedures. This is because the compilation process introduces impure features like cut and negation as failure. These are interpreted procedurally and their interpretation can depend on the order of literals and clauses. The compiler described below preserves the original order of expression nesting and tracts as much as possible, and this will determine the behaviour of the target Prolog procedure. Alternatively, one could use information about the intended mode to influence the order of clauses and literals. We see no way to avoid this problem of procedural non-equivalence as long as the target language contains impure features.

Here is the compiler.

1. First we write each tract into the form of a *program statement*, as defined by [Lloyd 87], p107, *i.e.* a formula of the form $A \leftarrow W$, where A is an atom and W is a first order formula.

   Unconditional and Conditional tracts are rewritten from the form:

   $$Head \leftrightarrow Body$$
   $$or$$
   $$Condition \rightarrow Head \leftrightarrow Body$$

   to the form:

   $$Head \leftarrow Body$$
   $$or$$
   $$Head \leftarrow Condition \wedge Body$$

   Note that these forms are procedurally, but not logically equivalent. Their procedural equivalence relies on the closed world assumption in Prolog. If it was known that *Head* was only ever to be called in fully ground mode then a cut could be safely inserted between the *Condition* and the *Body* of conditional statements, but this will not be safe in other modes.

---

[6]In as much as a Prolog procedure may be said to have a logical meaning.

For instance, under this transformation our $del/3$ procedure (tracts 4 and 5 above) becomes:

$$del(X, [], L) \leftarrow false$$
$$del(X, [Hd|Tl], L) \leftarrow (X = Hd \wedge Tl = L) \vee$$
$$(\exists L'\ L = [Hd|L'] \wedge del(X, Tl, L'))$$

2. Next we apply the Lloyd-Topor translation process to turn each program statement into a set of clauses. Essentially, this puts the program statement bodies into clausal form, but with some modifications to minimise the number of negated literals. Quantifiers are eliminated, negations are moved inwards and disjunctions cause a statement to split into two.

For instance, our $del/3$ procedure becomes:

$$del(X, [], L) \leftarrow false$$
$$del(X, [Hd|Tl], L) \leftarrow X = Hd \wedge Tl = L$$
$$del(X, [Hd|Tl], L) \leftarrow L = [Hd|L'] \wedge del(X, Tl, L')$$

3. Clauses can now be tidied up by removing equalities between variables and terms. This step is optional since it affects neither the meaning nor the behaviour of the clause.

Any literal in a clause body of the form $Variable = Term$ or $Term = Variable$ is omitted and each occurrence of $Variable$ in the clause is replaced by $Term$.

For instance, our $del/3$ procedure becomes:

$$del(X, [], L) \leftarrow false$$
$$del(Hd, [Hd|Tl], Tl) \leftarrow true$$
$$del(X, [Hd|Tl], [Hd|L']) \leftarrow del(X, Tl, L')$$

4. Each clause whose body contains the literal $false$ is deleted. Each literal $true$ is omitted from its clause body.

For instance, our $del/3$ procedure becomes:

$$del(Hd, [Hd|Tl], Tl)$$
$$del(X, [Hd|Tl], [Hd|L']) \leftarrow del(X, Tl, L')$$

5. Finally, we rewrite the clauses into standard Prolog notation. Each clause of the form:

$$Head \leftarrow Body_1 \wedge \ldots \wedge Body_n$$

is rewritten to the form:

$$Head :- Body_1, \ldots, Body_n.$$

and each body literal of the form $\neg Atom$ is replaced by $not\ Atom$.

For instance, our $del/3$ procedure becomes:

$$del(Hd, [Hd|Tl], Tl).$$
$$del(X, [Hd|Tl], [Hd|L']) :- del(X, Tl, L').$$

12   Conclusion

We have shown apply the 'proofs as programs' synthesis technique to logic programs. The key idea is to regard predicates as functions onto truth values in the 'all-ground' input mode, *i.e.* when all arguments are instantiated to ground terms. The synthesised procedure can then be run in other modes. In these other modes it will be defined for all inputs, but may give multiple outputs, and/or it may not terminate. That is, there will be at least one tract that matches any procedure call, but there may be many in non-all-ground modes and there may be an infinite number of computation routes.

Instead of synthesising Prolog programs directly, we synthesise procedures in a 'pure logic' form and then compile them into Prolog as a post-processing step. This is necessary in order to regard logic programs as functions onto truth values — these truth values are only implicit as 'success' or 'failure' in Prolog programs, and must be made explicit for the technique to work. This prevents us from reasoning with Prolog programs directly. It also prevents us from dealing with the non-declarative features of Prolog.

We have successfully tested this 'proofs as programs' technique on the synthesis of *del*/3 using the OYSTER system. In order to synthesise a first order program, it was necessary to avoid the use of any higher-order features of the OYSTER logic. To obviate this need to avoid higher-order features, we plan to build a synthesis system especially geared to the synthesis of pure logic programs. The first-order, Deductive Tableau System of [Manna & Waldinger 87] might be a better role model for this than Nuprl or OYSTER. However, it would also be interesting to use OYSTER, or a similar higher-order system, to explore the synthesis of higher-order logic programs.

In principle, it ought to be possible to use the 'proofs as programs' technique for the synthesis, transformation and verification of logic programs. We have not conducted large scale testing of our technique, so we do not yet have the empirical evidence to assess this potential. We plan to do this.

We also plan to compare our technique with other approaches to the transformation of logic programs. In particular, we want to compare it to systems that synthesise (or transform into) recursive programs, *e.g.* [Bruynooghe *et al* 89]. In our technique this necessarily requires proof by mathematical induction. We suspect that something like induction is also present in these other techniques, although it is sometimes disguised as 'loop spotting' in a symbolic execution tree.

References

[Bruynooghe *et al* 89]   M. Bruynooghe, D. de Schreye, and B. Krekels. Compiling control. *Journal of Logic Programming*, 135–162, 1989.

[Bundy 88a]   A. Bundy. A broader interpretation of logic in logic programming. In *Proceedings of the Fifth International Logic Programming Conference/ Fifth Symposium on Logic Programming*, pages 1624–1648, MIT Press, 1988. Also available from Edinburgh as Research Paper No. 388.

[Bundy 88b]   A. Bundy. *Proposal for a Recursive Techniques Editor for Prolog.*
                   Research Paper 394, Dept. of Artificial Intelligence, Edinburgh,
                   1988. Submitted to the special issue of Instructional Science on
                   Learning Prolog: Tools and Related Issues.

[Constable *et al* 86] R.L. Constable, S.F. Allen, H.M. Bromley, *et al. Implementing
                   Mathematics with the Nuprl Proof Development System.* Prentice
                   Hall, 1986.

[Hogger 81]   C.J. Hogger. Derivation of logic programs. *JACM*, 28(2):372–
                   392, April 1981.

[Horn 88]   C. Horn. *The Nurprl Proof Development System.* Working pa-
                   per 214, Dept. of Artificial Intelligence, Edinburgh, 1988. The
                   Edinburgh version of Nurprl has been renamed Oyster.

[Lloyd 87]   J.W. Lloyd. *Foundations of Logic Programs. Symbolic Compu-
                   tation*, Springer-Verlag, 1987. Second, extended edition.

[Manna & Waldinger 87] Z. Manna and R. Waldinger. The origin of a binary-search
                   paradigm. *Science of Computer Programming*, 9:37–83, 1987.

[Martin-Löf 79]   Per Martin-Löf. Constructive mathematics and computer pro-
                   gramming. In *6th International Congress for Logic, Methodology
                   and Philosophy of Science*, pages 153–175, Hanover, August 1979.
                   Published by North Holland, Amsterdam. 1982.

Studies in Pure Prolog: Termination *

Krzysztof R. Apt
Centre for Mathematics and Computer Science
Kruislaan 413, 1098 SJ Amsterdam, The Netherlands

Dino Pedreschi
Dipartimento di Informatica, Università di Pisa
Corso Italia 40, 56125 Pisa, Italy

Abstract

We provide a theoretical basis for studying termination of logic programs with
the Prolog selection rule. To this end we study the class of *left terminating* programs. These are logic programs that terminate with the Prolog selection rule for
all ground goals. First we show that various ways of defining semantics coincide
for left terminating programs. Then we offer a characterization of left terminating
programs that provides us with a practical method of proving termination. The
method is proven to be complete and is illustrated by giving simple proofs of termination of the quicksort, permutation and mergesort programs for the desired class
of goals.

1   Introduction

Background

Algorithms are designed for two types of problems - decidable ones and semi-decidable
ones. In the latter case we cannot claim termination for all inputs. In the former case
we usually can and only in few cases - like interactive programs (game playing programs,
editors, ...) or operating systems, we choose not to do so.

In this paper we study termination of Prolog programs and, naturally, confine our
attention to the category of programs that terminate for all inputs. By termination
we mean here finiteness of *all* possible Prolog derivations starting in the initial goal.
However, in the case of Prolog programs one is confronted with the problem that an
apparently correct program may fail to terminate in this sense for certain forms of inputs.

---

*This research was partly done during the authors' stay at the Department of Computer Sciences,
University of Texas at Austin, Austin, Texas, U.S.A. . First author's work was partly supported by
ESPRIT Basic Research Action 3020 (Integration). Second author's work was partly supported by
ESPRIT Basic Research Action 3012 (Compulog) and by the Italian National Research Council – C.N.R..

For example, the append program fails to terminate in this sense for a goal with all arguments being variables. To cope with this complication we only require that the program terminates for all *ground* inputs. In such cases only "yes" or "no" answer can be given. We call such programs *left terminating*. Then to show that a Prolog program exhibits a proper termination behaviour we first show that it is left terminating and then that it terminates for certain types of non-ground inputs. Our method of showing the former will also allow us to establish the latter.

When studying Prolog programs from the point of view of termination it is useful to notice that some programs terminate for all ground goals for *all* selection rules. Such programs are extensively studied in Bezem [Bez89] where they are called *terminating programs*. These are usually programs whose termination depends on a simple reduction in one or more arguments. Examples of terminating programs are append, member, N queens, various tree insertion and deletion programs and several others.

However, some Prolog programs satisfy such a strong termination property but fail to terminate for certain desired forms of inputs for some selection rules.

An example is the following append3 program in which the append program is used:

```
append3(Xs, Ys, Zs, Us) ←
    append(Xs, Ys, Vs),
    append(Vs, Zs, Us).
```

Then append3 is a terminating program which terminates for the goal ← append3(xs, ys, zs, Us), where xs, ys, zs are lists and Us a variable, when the Prolog selection rule is used but fails to terminate when the rightmost selection rule is used.

Worse yet, some programs fail to be terminating even though they terminate for the Prolog selection rule for the desired class of inputs. An example is the flatten program which collects all the nodes of a tree in a list:

```
flatten(nil, []) ←.
flatten(t(L, X, R), Xs) ←
    flatten(L, X1s),
    flatten(R, X2s),
    append(X1s, [X | X2s], Xs).
```

flatten is not a terminating program but it terminates for the goal ← flatten(x, Xs), where x is a ground term and Xs a variable, when the Prolog selection rule is used.

In general, the problem arises due to the use of local variables, i.e. variables which appear in the body of a clause but not in its head. Several left terminating Prolog programs use local variables in an essential way and consequently fail to be terminating. Examples of such programs are various sorting and permutation programs and graph searching programs. Programs which fall into this category are usually of the form "generate and test" or "divide and conquer".

In this paper we provide a framework to study left terminating programs. To this end we refine the ideas of Bezem [Bez89] and Cavedon [Cav89] and use the concept of a level mapping. This is a function assigning natural numbers to ground atoms. Our main tool is the concept of an *acceptable program*. Intuitively, a program is acceptable if for some level mapping, for all ground instances of the clauses of the program, the level of the head is smaller than the level of atoms in a certain prefix of the body. Which prefix is considered is determined by some model of the program.

The main result of the paper is that the notions of left termination and acceptability coincide. The proof of this fact uses an iterated multiset ordering. This equivalence result provides us with a method of proving left termination. Moreover, it allows us to prove termination of a left terminating Prolog program for a class of non-ground goals. The method is easy to use and is illustrated by proving termination of the quicksort, permutation and mergesort programs.

Plan of the paper

This paper is organized as follows. In the next section we introduce the concept of a left terminating program. This is a program that terminates for all ground goals w.r.t. Prolog selection rule. We show that left terminating programs satisfy an elegant semantic property: the least Herbrand model of a left terminating program P is a unique fixpoint of the immediate consequence operator T_P associated with P, can be identified with the unique fixpoint of the 3-valued immediate consequence operator associated with P and can be characterized in terms of the completion of P, $comp(P)$.

In Section 3 we provide a useful characterization of left terminating programs by introducing the notion of an acceptable program and proving that the notions of acceptability and left termination coincide. The crucial concept here is that of a bounded goal. It allows us to characterize terminating goals.

Finally, in Section 4 we prove left termination of the quicksort, permutation and mergesort programs by providing in each case a simple proof of acceptability. Using the concept of boundedness we show that each program terminates w.r.t. a desired class of non-ground goals.

Preliminaries

We use standard notation and terminology of Lloyd [Llo87] or Apt [Apt88]. In particular, we use the following abbreviations for a logic program P (or simply a *program*):

B_P for the Herbrand Base of P,

T_P for the immediate consequence operator of P,

M_P for the least Herbrand model of P,

$ground(P)$ for the set of all ground instances of clauses from P,

$comp(P)$ for Clark's completion of P.

Also, we use Prolog's convention identifying in the context of a program each string starting with a capital letter with a variable, reserving other strings for the names of constants, terms or relations. So, for example Xs stands for a variable whereas xs stands for a term.

In the programs we use the usual list notation. The constant [ ] denotes the empty list and [. | .] is a binary function which given a term x and a list xs produces a new list $[x \mid xs]$ with head x and tail xs. By convention, identifiers ending with "s", like xs, will range over lists. The standard notation $[x_1, \ldots, x_n]$, for $n \geq 0$, is used as an abbreviation of $[\, x_1 \mid [\ldots [\, x_n |[\,]\,]\ldots]\,]$. In general, the Herbrand Universe will also contain "impure" elements that contain $[\,]$ or $[\,.\mid.\,]$ but are not lists - for example $s([\,])$ or $[s(0)\mid 0]$ where 0 is a constant and s a unary function symbol. They will not cause any complications.

Given an operator T on a complete partial ordering L with the least element $\bot$, we define the *upward ordinal powers* of T starting at $\bot$ in the standard way and denote them by $T \uparrow \alpha$ where α is an ordinal. If L has the greatest element, say $\top$, (this is the case when for example L is a complete lattice) we define the *downward ordinal powers* of T starting at $\top$ in the standard way and denote them by $T \downarrow \alpha$.

Throughout the paper we consider SLD-resolution with one selection rule only – namely that of Prolog, usually called the leftmost selection rule. As S in SLD stands for "selection rule", we denote this form of resolution by LD (*Linear resolution for Definite clauses*). The concepts of LD-derivation, LD-refutation, LD-tree, etc. are then defined in the usual way. By "pure Prolog" we mean in this paper the LD-resolution combined with the depth first search in the LD-trees.

By choosing variables of the input clauses and the used mgu's in a fixed way we can assume that for every program P and goal G there exists exactly one LD-tree for $P \cup \{G\}$.

2 Left Termination

Our interest here is in terminating Prolog programs. This motivates the following concept.

Definition 2.1 A program P is called *left terminating* if all LD-derivations of P starting in a ground goal are finite. □

In other words, a program is left terminating if all LD-trees for P with a ground root are finite. When studying Prolog programs, one is actually interested in proving termination of a given program not only for all ground goals but also for a class of non-ground goals constituting the intended queries. Our method of proving left termination will allow us to identify for each program such a class of non-ground goals.

But first let us see some simple consequences of the above definition. Following Blair [Bla86] a program is called *determinate* if $T_P \uparrow \omega = T_P \downarrow \omega$.

Theorem 2.2 *Every left terminating program is determinate.*

Proof. By the results of Apt and Van Emden [AvE82] (see also Lloyd [Llo87]) for every program P

$T_P \uparrow \omega = \{A \in B_P \mid \text{there exists a successful } SLD\text{-tree for } P \cup \{\leftarrow A\}\}$,
$T_P \downarrow \omega = \{A \in B_P \mid \text{there does not exist a finitely failed } SLD\text{-tree for } P \cup \{\leftarrow A\}\}$.

We always have $T_P \uparrow \omega \subseteq T_P \downarrow \omega$, since T_P is monotonic. To prove the converse inclusion for a left terminating program P, take some $A \in T_P \downarrow \omega$. By the second equality the

LD-tree for $P \cup \{\leftarrow A\}$ is not finitely failed. But by the choice of P it is finite, so it is successful. Thus by the first equality $A \in T_P \uparrow \omega$.

$\square$

The converse of the above theorem does not hold - it suffices to take $P = \{A \leftarrow A, B\}$. Then $T_P \uparrow \omega = \emptyset$ and $T_P \downarrow \omega = \emptyset$ but P is not left terminating.

The determinate programs, and consequently left terminating programs, enjoy some pleasing semantic properties it is useful to record.

Theorem 2.3 *For a determinate program P, M_P is the unique fixpoint of T_P.*

Proof. We prefer to give a more general proof of this fact. To this end consider a monotonic operator T on a complete lattice. Then by monotonicity
(i) for every fixpoint Y of T

$$T \uparrow \omega \subseteq Y \subseteq T \downarrow \omega,$$

(ii) $T \uparrow \omega \subseteq T \uparrow (\omega + 1) \subseteq T \downarrow \omega$.

Suppose now that $T \uparrow \omega = T \downarrow \omega$. Then by (i) T has at most one fixpoint and by (ii) $T \uparrow \omega$ is a fixpoint of T, since by definition $T \uparrow (\omega + 1) = T(T \uparrow \omega)$.
The claim of the theorem now follows, since T_P is monotonic and by the result of Apt and Van Emden [AvE82] $M_P = T_P \uparrow \omega$.

$\square$

The other property of determinate programs is based on the theory of 3-valued models for logic programs developed by Fitting [Fit85]. We recall first the relevant definitions and results. Fitting [Fit85] uses a 3-valued logic due to Kleene [Kle52].

In Kleene's logic there are three truth values: **t** for true, **f** for false and **u** for undefined. Every connective takes the value **t** or **f** if it takes that value in 2-valued logic for all possible replacements of **u**'s by **t** or **f**; otherwise it takes value **u**.

A Herbrand interpretation for this logic (called a 3-*valued* Herbrand interpretation) is defined as a pair (T, F) of disjoint sets of ground atoms. Given such an interpretation $I = (T, F)$ a ground atom A is true in I if $A \in T$, false in I if $A \in F$ and undefined otherwise. Given $I = (T, F)$ we denote T by I^+ and F by I^-. Thus $I = (I^+, I^-)$. If $I^+ \cup I^- = B_P$, we call I a *total* 3-valued Herbrand interpretation for the program P.

Every (2-valued) Herbrand interpretation I for a program P determines a total 3-valued Herbrand interpretation $(I, B_P - I)$ for P. This allows us to identify every 2-valued Herbrand interpretation I with its 3-valued counterpart $(I, B_P - I)$.

Given a program P, the 3-valued Herbrand interpretations for P form a complete partial ordering with the ordering $\subseteq$ defined by

$$I \subseteq J \text{ iff } I^+ \subseteq J^+ \wedge I^- \subseteq J^-$$

and with the least element $(\emptyset, \emptyset)$. Note that in this ordering every total 3-valued Herbrand interpretation is $\subseteq$-maximal.

Following Fitting [Fit85], given a program P we define an operator Φ_P on the complete partial ordering of 3-valued Herbrand interpretations for P as follows:

$$\Phi_P(I) = (T, F),$$

where

$T = \{A \mid \text{there exists } A \leftarrow B_1, \ldots, B_k \text{ in } ground(P) \text{ with } B_1 \wedge \ldots \wedge B_k \text{ true in } I\},$
$F = \{A \mid \text{for all } A \leftarrow B_1, \ldots, B_k \text{ in } ground(P), B_1 \wedge \ldots \wedge B_k \text{ is false in } I\}.$

It is easy to see that T and F are disjoint, so $\Phi_P(I)$ is indeed a 3-valued Herbrand interpretation. Φ_P is a natural generalization of the operator T_P to the case of 3-valued logic. Φ_P is easily seen to be monotonic. The following observation of Fitting [Fit85] is of relevance here.

Lemma 2.4 *For every program P and ordinal α*

$$\Phi_P \uparrow \alpha = (T_P \uparrow \alpha, \; B_P - T_P \downarrow \alpha).$$

$\square$

This implies the following results.

Lemma 2.5 *For a determinate program P, $M_P = \Phi_P \uparrow \omega$.* $\square$

Proof. By Lemma 2.4 and the fact that $M_P = T_P \uparrow \omega$.

Corollary 2.6 *For a determinate program P, M_P is the unique fixpoint of Φ_P.*

Proof. Let Y be a fixpoint of Φ_P. By the monotonicity of Φ_P, $\Phi_P \uparrow \omega \subseteq Y$, so by Lemma 2.5, $M_P \subseteq Y$. But M_P is a total 3-valued Herbrand interpretation so it is $\subseteq$-maximal and consequently $M_P = Y$. $\square$

The final characterization of the model M_P for determinate programs is in terms of the completion $comp(P)$.

Theorem 2.7 *For a determinate program P, for all ground atoms $A \in B_P$*

$$\begin{aligned}
M_P &\models A & &\text{iff } comp(P) \models A, \\
M_P &\models \neg A & &\text{iff } comp(P) \models \neg A.
\end{aligned}$$

Proof. Combining various completeness and characterization results (see Lloyd [Llo87] or Apt [Apt88]) we have for every logic program P,

$$\begin{aligned}
T_P \uparrow \omega &\models A & &\text{iff } comp(P) \models A, \\
T_P \downarrow \omega &\models \neg A & &\text{iff } comp(P) \models \neg A.
\end{aligned}$$

But for a determinate program P, $M_P = T_P \uparrow \omega = T_P \downarrow \omega$.   □

Corollary 2.8 *For a determinate program* P

$$M_P = \{A \in B_P \mid comp(P) \models A\},$$
$$M_P = \{A \in B_P \mid comp(P) \not\models \neg A\}.$$

□

Thus for determinate programs, and a fortiori for left terminating programs, three most common approaches to semantics coincide and result in a simple declarative semantics in the form of a unique fixpoint of the T_P operator which coincides with the unique fixpoint of the Φ_P operator and which can be characterized by means of the completion $comp(P)$.

3 Proving Left Termination

Let us consider now how to prove that a program is left terminating. Starting from Floyd [Flo67] the classical proofs of program termination have been based on the use of well-founded orderings. This approach has been successfully used in the area of logic programming (see e.g. Bezem [Bez89], Cavedon [Cav89]) but with no attention paid to Prolog programs. The notable exception is Deville [Dev90].

We obtain the desired method by a modification of the ideas of Bezem [Bez89] and Cavedon [Cav89].

Recurrent Programs

It is useful to recall first some concepts and results from Bezem [Bez89]. A *level mapping* for a program P is a function $| \; | : B_P \to N$ of ground atoms to natural numbers. For $A \in B_P$, $|A|$ is the level of A. Following Bezem [Bez89] (see also Cavedon [Cav89]), a program is called *recurrent* if for some level mapping $| \; |$, for every clause $A \leftarrow B_1, \ldots, B_n$ in $ground(P)$

$$|A| > |B_i| \text{ for } i \in [1, n].$$

Another relevant concept is that of *boundedness*: an atom A is bounded with respect to a level mapping $| \; |$ if $| \; |$ is bounded on the set $[A]$ of ground instances of A. A goal is bounded if all its atoms are. Bezem [Bez89] showed that every SLD-derivation of a recurrent program starting in a bounded goal terminates.

A program is called *terminating*, if all its SLD-derivations starting in a ground goal are finite. Hence, terminating programs have the property that the SLD-trees of ground goals are finite, and any search procedure in such trees will always terminate, independently from the adopted selection rule.

One of the main results in Bezem [Bez89] is that a program is recurrent if and only if it is terminating. Because of this result recurrent programs and bounded goals are too restrictive concepts to deal with Prolog programs, as a larger class of programs and goals is terminating when adopting a specific selection rule, e.g. Prolog selection rule.

Example 3.1

(i) Consider the following program **even** which defines even numbers and the "less than or equal" relation:

```
even(0) ←.
even(s(s(X))) ← even(X).

lte(0,Y) ←.
lte(s(X),s(Y)) ← lte(X,Y).
```

even is recurrent with $|even(s^n(0))| = n$ and $|lte(s^n(0), s^m(0))| = \min\{n, m\}$. Now consider the goal:

$$G =\leftarrow lte(x, s^{100}(0)), even(x)$$

which is supposed to compute the even numbers not exceeding 100. The LD-tree for G is finite, whereas there exists an infinite SLD-derivation when the rightmost selection rule is used. As a consequence of Bezem's result, the goal G is not bounded, although it can be evaluated by a finite Prolog computation.

Actually, most "generate and test" Prolog programs are not recurrent, as they heavily depend on the left-to-right order of evaluation, like the example above.

(ii) Consider the following naive **reverse** program:

```
reverse([], []) ←.
reverse([X | Xs], Ys) ←
  reverse(Xs, Zs),
  append(Zs, [X], Ys).

append([], Ys, Ys) ←.
append([X | Xs], Ys, [X | Zs]) ← append(Xs, Ys, Zs).
```

The ground goal $\leftarrow reverse(xs, ys)$, for arbitrary lists xs and ys, has an infinite SLD-derivation, obtained by using the selection rule which selects the leftmost atom at the first two steps, and the second leftmost atom afterwards. By Bezem's result, **reverse** is not recurrent.

(iii) Consider the following program DC, representing a (binary) "divide and conquer" schema; it is parametric with respect to the *base, conquer, divide* and *merge* predicates.

```
dc(X,Y) ←
  base(X),
  conquer(X,Y).
```

```
dc(X,Y) ←
   divide(X,X1,X2),
   dc(X1,Y1),
   dc(X2,Y2),
   merge(Y1,Y2,Y).
```

Many programs naturally fit into this schema, or its generalization to non fixed arity of the divide/merge predicates. Unfortunately, DC is not recurrent: it suffices to take a ground instance of the recursive clause with $X = a$, $X1 = a$, $Y = b$, $Y1 = b$, and observe that the atom $dc(a,b)$ occurs both in the head and in the body of such a clause. In this example, the leftmost selection rule is needed to guarantee that the input data is divided into subcomponents before recurring on such subcomponents. □

Acceptable Programs

To cope with these difficulties we modify the definition of a recurrent program as follows.

Definition 3.2 Let P be a program, $||$ a level mapping for P and I a (not necessarily Herbrand) model of P. P is called *acceptable with respect to* $||$ *and* I if for every clause $A \leftarrow B_1, \ldots, B_n$ in $ground(P)$

$$|A| > |B_i| \text{ for } i \in [1, \bar{n}],$$

where

$$\bar{n} = \min(\{n\} \cup \{i \in [1, n] \mid I \not\models B_i\}).$$

Alternatively, we may define $\bar{n}$ by

$$\bar{n} = \begin{cases} n & \text{if } I \models B_1 \wedge \ldots \wedge B_n, \\ i & \text{if } I \models B_1 \wedge \ldots \wedge B_{i-1} \text{ and } I \not\models B_1 \wedge \cdots \wedge B_i. \end{cases}$$

P is called *acceptable* if it is acceptable with respect to some level mapping and a model of P. □

Thus, given a level mapping $||$ for P and a model I of P, in the definition of acceptability w.r.t. $||$ and I for every clause $A \leftarrow B_1, \ldots, B_n$ in $ground(P)$ we only require that the level of A is higher than the level of B_i's in a certain prefix of $B_1, \ldots, B_n$. Which B_i's are taken into account is determined by the model I. If $I \models B_1 \wedge \ldots \wedge B_n$ then all of them are considered and otherwise only those whose index is $\leq \bar{n}$, where $\bar{n}$ is the least index i for which $I \not\models B_i$.

The idea underlying the above definition can be illustrated by the following example. Consider a program P containing the clause

$$p(X) \leftarrow q(X, Y), r(Y)$$

and a model I of P. Consider two ground instances

$$(c_1) \quad p(a) \leftarrow q(a,b), r(b),$$
$$(c_2) \quad p(a) \leftarrow q(a,c), r(c)$$

of this clause (assuming that the constants a, b, c are in the Herbrand Universe of P) and suppose that $q(a,b) \in I$ but $q(a,c) \notin I$. To prove acceptability, a level mapping $|\,|$ is supposed to satisfy

$$|p(a)| > |q(a,b)| \text{ and } |p(a)| > |r(b)|$$

for clause (c_1), but only

$$|p(a)| > |q(a,c)|$$

for clause (c_2). Intuitively, the condition $q(a,c) \notin I$ excludes (by the soundness of the SLD-resolution) the existence of a refutation for $q(a,c)$ and consequently there is no point in checking that the level mapping decreases from $p(a)$ to $r(c)$, since the Prolog interpreter will never reach $r(c)$ during the execution starting with the goal $\leftarrow p(a)$.

The following observation is immediate.

Lemma 3.3 *Every recurrent program is acceptable.*

Proof. Take $I = B_P$. Then for every $A \leftarrow B_1, \ldots, B_n$ in $ground(P)$, $\bar{n} = n$. $\square$

Our aim is to prove that the notions of acceptability and left termination coincide.

Multiset ordering

To prove one half of this statement we use the multiset ordering. A *multiset*, sometimes called *bag*, is an unordered sequence. Given a (non-reflexive) ordering $<$ on a set W, the *multiset ordering over* $(W, <)$ is an ordering on finite multisets of the set W. It is defined as the transitive closure of the relation in which X is smaller than Y if X can be obtained from Y by replacing an element a of Y by a finite (possibly empty) multiset each of whose elements is smaller than a in the ordering $<$.

In symbols, first we define the relation $\prec$ by

$$X \prec Y \text{ iff } X = Y - \{a\} \cup Z \text{ for some } Z \text{ such that } b < a \text{ for } b \in Z,$$

where X, Y, Z are finite multisets of elements of W, and then define the multiset ordering over $(W, <)$ as the transitive closure of the relation $\prec$.

It is well-known (see e.g. Dershowitz [Der87]) that multiset ordering over a well-founded ordering is again well-founded. Thus it can be iterated while maintaining well-foundedness.

What we need in our case is two fold iteration. We start with the set of natural numbers N ordered by $<$ and apply the multiset ordering twice. We call the first iteration multiset ordering and the second *double multiset ordering*. Both are well-founded. The double multiset ordering is defined on the finite *multisets* of finite multisets of natural numbers, but we shall use it only on the finite *sets* of finite multisets of natural numbers. The following lemma will be of help when using the double multiset ordering.

Lemma 3.4 *Let X and Y be two finite sets of finite multisets of natural numbers. Suppose that*

$$\forall x \in X \, \exists y \in Y \, (y \, majorizes \, x),$$

where y majorizes x means that x is smaller than y in the multiset ordering.

Then X is smaller than Y in the double multiset ordering.

Proof. We call an element $y \in Y$ *majorizing* if it majorizes some $x \in X$. X can be obtained from Y by first replacing each majorizing $y \in Y$ by the multiset M_y of elements of X it majorizes and then removing from Y the non-majorizing elements. This proves the claim. □

Below we use the notation bag $(a_1, \ldots, a_n)$ to denote the multiset consisting of the unordered sequence $a_1, \ldots, a_n$.

Boundedness

Another important concept is boundedness. It allows us to identify goals from which no divergence can arise. Recall that an atom A is called *bounded* w.r.t. a level mapping $|\,|$ if $|\,|$ is bounded on the set $[A]$ of ground instances of A. If A is bounded, then $||[A]||$ denotes the maximum that $|\,|$ takes on $[A]$. Note that every ground atom is bounded.

Our concept of a bounded goal differs from that of Bezem [Bez89] in that it takes into account the model I. This results in a more complicated definition.

Definition 3.5 Let P be a program, $|\,|$ a level mapping for P, I a model of P and $k \geq 0$.

(i) With each ground goal $G = \leftarrow A_1, \ldots, A_n$ we associate a finite multiset $|G|_I$ of natural numbers defined by

$$|G|_I = \text{bag}\,(|A_1|, \ldots, |A_{\bar{n}}|),$$

where

$$\bar{n} = \min(\{n\} \cup \{i \in [1, n] \mid I \not\models A_i\}).$$

(ii) With each goal G we associate a set of multisets $||G||_I$ defined by

$$||G||_I = \{|G'|_I \mid G' \text{ is a ground instance of } G\}.$$

(iii) A goal G is called *bounded by k* w.r.t. $|\,|$ and I if $k \geq \ell$ for $\ell \in \cup ||G||_I$.

A goal is called *bounded* w.r.t. $|\,|$ and I if it is bounded by some $k \geq 0$ w.r.t. $|\,|$ and I.

□

It is useful to note the following.

Lemma 3.6 *Let P be a program, $|\,|$ a level mapping for P and I a model of P. A goal G is bounded w.r.t. $|\,|$ and I iff the set $||G||_I$ is finite.*

Proof. Consider a goal G that is bounded by some k. Suppose that G has n atoms. Then each element of $\|[G]\|_I$ is a multiset of at most n numbers selected from $[0, k]$. The number of such multisets is finite.

The other implication is obvious. $\qquad\square$

The following lemma is an analogue of Lemma 2.5 of Bezem [Bez89].

Lemma 3.7 *Let P be a program that is acceptable w.r.t. a level mapping $|\,|$ and a model I. Let G be a goal that is bounded (w.r.t. $|\,|$ and I) and let H be an LD-resolvent of G from P. Then*

(i) H is bounded,

(ii) $\|[H]\|_I$ is smaller than $\|[G]\|_I$ in the double multiset ordering.

Proof. Let $G = \leftarrow A_1, \ldots, A_n (n \geq 1)$. For some input clause $C = A \leftarrow B_1, \ldots, B_k (k \geq 0)$ and mgu θ of A and A_1, $H = \leftarrow (B_1, \ldots, B_k, A_2, \ldots, A_n)\theta$.

First we show that for every ground instance H_0 of H there exists a ground instance G' of G such that $|H_0|_I$ is smaller that $|G'|_I$ in the multiset ordering.

So let H_0 be a ground instance of H. For some substitution δ

$$H_0 = \leftarrow B_1', \ldots, B_k', A_2', \ldots, A_n'$$

and A_1' is ground, where for brevity for any atom, clause or goal B, B' denotes $B\theta\delta$. Note that

$$C' = A_1' \leftarrow B_1', \ldots, B_k'$$

and

$$G' = \leftarrow A_1', \ldots, A_n',$$

since $A' = A_1'$ as $A\theta = A_1\theta$.

Case 1 For $i \in [1, k]$ $I \not\models B_i'$.
Then
$$|H_0|_I = \text{bag}\,(|B_1'|, \ldots, |B_k'|, |A_2'|, \ldots, |A_{\bar{n}}'|)$$

where
$$\bar{n} = \min(\{n\} \cup \{i \in [2, n] \mid \not\models A_i'\}).$$

Additionally, since I is a model of P, $I \models A_1'$. Thus

$$|G'|_I = \text{bag}\,(|A_1'|, |A_2'|, \ldots, |A_{\bar{n}}'|).$$

This means that $|H_0|_I$ is obtained from $|G'|_I$ by replacing $|A_1'|$ by $|B_1'|, \ldots, |B_k'|$. But by the definition of acceptability
$$|B_i'| < |A_1'|$$

for $i \in [1, k]$, so $|H_0|_I$ is smaller than $|G'|_I$ in the multiset ordering. □

Case 2 For some $i \in [1, k]$ $I \not\models B'_i$.
Then

$$|H_0|_I = \text{bag}(|B'_1|, \ldots, |B'_{\bar{k}}|)$$

where

$$\bar{k} = \min(\{i \in [1, k] \mid I \not\models B'_i\}).$$

Also by the definition of acceptability

$$|B'_i| < |A'_1|$$

for $i \in [1, \bar{k}]$, so $|H_0|_I$ is smaller than $|G'|_I$ in the multiset ordering. □

This implies claim (i) since G is bounded. By Lemma 3.6 $|[H]|_I$ is finite and claim (ii) now follows by Lemma 3.4. □

Corollary 3.8 *Let P be an acceptable program and G a bounded goal. Then all LD-derivations of $P \cup \{G\}$ are finite.*

Proof. The double multiset ordering is well-founded. □

Corollary 3.9 *Every acceptable program is left terminating.*

Proof. Every ground goal is bounded. □

LD-trees

To prove the converse of Corollary 3.9 we analyze the size of finite LD-trees. To this end we need the following lemma, where $nodes_P(G)$ for a program P and a goal G denotes the number of nodes in the LD-tree for $P \cup \{G\}$.

Lemma 3.10 *(LD-tree) Let P be a program and G a goal such that the LD-tree for $P \cup \{G\}$ is finite. Then*

(i) for all substitutions θ, $nodes_P(G\theta) \leq nodes_P(G)$,

(ii) for all prefixes H of G, $nodes_P(H) \leq nodes_P(G)$,

(iii) for all non-root nodes H in the LD-tree for $P \cup \{G\}$, $nodes_P(H) < nodes_P(G)$.

Proof. (i) By an application of a variant of the Lifting Lemma (see e.g. Lloyd [Llo87]) to LD-derivations we conclude that to every LD-derivation of $P \cup \{G\theta\}$ with input clauses

$C_1, C_2, \ldots$, there corresponds an LD-derivation of $P \cup \{G\}$ with input clauses $C_1, C_2, \ldots$ of the same of larger length. This implies the claim.

(ii) Consider a prefix $H = \leftarrow A_1, \ldots, A_k$ of $G = \leftarrow A_1, \ldots, A_n$ $(n \geq k)$. By an appropriate renaming of variables (formally justified by the Variant Lemma 2.8 in Apt [Apt88]) we can assume that all input clauses used in the LD-tree for $P \cup \{H\}$ have no variables in common with G. We can now transform the LD-tree for $P \cup \{H\}$ into an initial subtree of the LD-tree for $P \cup \{G\}$ by replacing in it a node $\leftarrow B_1, \ldots, B_l$ by $\leftarrow B_1, \ldots, B_l, A_{k+1}\theta, \ldots, A_n\theta$, where θ is the composition of the mgu's used on the path from the root H to the node $\leftarrow B_1, \ldots, B_l$. This implies the claim.

(iii) Immediate by the definition. □

As stated at the beginning of Section 2, we are interested in proving not only left termination of a program, but also its termination for a class of non-ground goals. We now show that the concepts of acceptability and boundedness provide us with a complete method for proving both properties.

Theorem 3.11 *Let P be a left terminating program. Then for some level mapping $| \; |$ and a model I of P*

(i) P is acceptable w.r.t. $| \; |$ and I,

(ii) for every goal G, G is bounded w.r.t. $| \; |$ and I iff all LD-derivations of $P \cup \{G\}$ are finite.

Proof. Define the level mapping by putting for $A \in B_P$

$$|A| = \text{nodes}_P \, (\leftarrow A).$$

Since P is left terminating, this level mapping is well defined. Next, choose

$$I = \{A \in B_P \,|\, \text{there is an } LD\text{-refutation of } P \cup \{\leftarrow A\}\}.$$

By the strong completeness of SLD-resolution, $I = M_P$, so I is a model of P.

First we prove one implication of (ii).

(ii1) Consider a goal G such that all LD-derivations of $P \cup \{G\}$ are finite. We prove that G is bounded by $\text{nodes}_P(G)$ w.r.t. $| \; |$ and I.

To this end take $\ell \in \cup ||[G]||_I$. For some ground instance $\leftarrow A_1, \ldots, A_n$ of G and $i \in [1, \bar{n}]$, where

$$\bar{n} = \min(\{n\} \cup \{i \in [1, n] \,|\, I \not\models A_i\}),$$

we have $\ell = |A_i|$. We now calculate

$$\text{nodes}_P(G)$$
$$\geq \quad \{\text{Lemma 3.10 (i)}\}$$

$$nodes_P(\leftarrow A_1, \ldots, A_n)$$
$\geq$ {Lemma 3.10 (ii)}
$$nodes_P(\leftarrow A_1, \ldots, A_{\bar{n}})$$
$\geq$ {Lemma 3.10 (iii), noting that for $j \in [1, \bar{n}-1]$
there is an LD-refutation of $P \cup \{\leftarrow A_1, \ldots, A_j\}$}
$$nodes_P(\leftarrow A_i, \ldots, A_{\bar{n}})$$
$\geq$ {Lemma 3.10 (ii)}
$$nodes_P (\leftarrow A_i)$$
$=$ {definition of $|\ |$}
$$|A_i|$$
$=$ ℓ.

(i) We now prove that P is acceptable w.r.t. $|\ |$ and I. Take a clause $A \leftarrow B_1, \ldots, B_n$ in P and its ground instance $A\theta \leftarrow B_1\theta, \ldots, B_n\theta$. We need to show that

$$|A\theta| > |B_i\theta| \text{ for } i \in [1, \bar{n}],$$

where

$$\bar{n} = \min(\{n\} \cup \{i \in [1, n] \mid I \not\models B_i\theta\}).$$

We have $A\theta\theta \equiv A\theta$, so $A\theta$ and A unify. Let $\mu = mgu(A\theta, A)$. Then $\theta = \mu\delta$ for some δ. By the definition of LD-resolution, $\leftarrow B_1\mu, \ldots, B_n\mu$ is an LD-resolvent of $\leftarrow A\theta$. Then for $i \in [1, \bar{n}]$

$$|A\theta|$$
$=$ {definition of $|\ |$}
$$nodes_P (\leftarrow A\theta)$$
$>$ {Lemma 3.10 (iii), $\leftarrow B_1\mu, \ldots, B_n\mu$ is a resolvent of $\leftarrow A\theta$}
$$nodes_P (\leftarrow B_1\mu, \ldots, B_n\mu)$$
$\geq$ {part (ii1), noting that $B_i\theta \in \cup|[\leftarrow B_1\mu, \ldots, B_n\mu]|_I$}
$$|B_i\theta|.$$

(ii2) Consider a goal G which is bounded w.r.t. $|\ |$ and I. Then by (i) and Corollary 3.8 all LD-derivations of $P \cup \{G\}$ are finite. $\qquad\square$

Corollary 3.12 A program is left terminating iff it is acceptable.

Proof. By Corollary 3.9 and Theorem 3.11. $\qquad\square$

4   Applications

The equivalence between the left terminating and acceptable programs provides us with a method of proving termination of Prolog programs. The level mapping and the model used in the proof of Theorem 3.11 were quite involved and relied on elaborate information about the program at hand which is usually not readily available. However, in practical situations much simpler constructions suffice. The level mapping can be usually defined as a simple function of the terms of the ground atom and the model takes into account only some straightforward information about the program. We illustrate it by means of three examples.

First, we define by structural induction a function $|\,|$ on ground terms by putting:

$$||[x|xs]|| = |xs| + 1,$$
$$|f(x_1,\ldots,x_n)| = 0 \text{ if } f \neq [.\,|\,.].$$

It is useful to note that for a list xs, $|xs|$ equals its length. The function $|\,|$ is called *listsize* in Ullman and Van Gelder [UvG88]. It will be used in the examples below.

Quicksort

Consider the following program QS (for quicksort):

(qs_1) qs([], []) ←.
(qs_2) qs([X | Xs], Ys) ←
          f(X, Xs, X1s, X2s),
          qs(X1s, Y1s),
          qs(X2s, Y2s),
          a(Y1s, [X | Y2s], Ys).

(f_1) f(X, [], [], []) ←.
(f_2) f(X, [Y | Xs], [Y | Y1s], Y2s) ←
          X > Y,
          f(X, Xs, Y1s, Y2s).
(f_3) f(X, [Y | Xs], Y1s, [Y | Y2s]) ←
          X ≤ Y,
          f(X, Xs, Y1s, Y2s).

(a_1) a([], Ys, Ys) ←.
(a_2) a([X | Xs], Ys, [X | Zs]) ←
          a(Xs, Ys, Zs).

We assume that QS operates on the domain of natural numbers over which the builtin relations $>$ and $\leq$, written in infix notation, are defined. This domain can be incorporated into the Herbrand universe of QS by adding to the language of QS the constant 0 and the successor function s (for example by adding to QS the clause s(0) > 0 ←.).

Denote now the program consisting of the clauses $(f_1), (f_2), (f_3)$ by `filter`, and the program consisting of the clauses $(a_1), (a_2)$ by `append`.

Lemma 4.1 `filter` *is recurrent with* $|f(x, xs, x1s, x2s)| = |xs|$. □

We adopted here the simplifying assumption that builtins $>$ and $\leq$ are recurrent with the level mapping $|s > t| = 0$ and $|s \leq t| = 0$.

Lemma 4.2 `append` *is recurrent with* $|a(xs, ys, zs)| = |xs|$. □

Lemma 4.3 `QS` *is not recurrent.*

Proof. Consider clause (qs_2) instantiated with the ground substitution

$$\{X/a,\ Xs/b,\ Ys/c,\ X1s/[a|b],\ Y1s/c\}.$$

Then the ground atom $qs([a|b], c)$ appears both in the head and the body of the resulting clause. □

To prove that `QS` is left terminating we show that it is acceptable. We define an appropriate level mapping $|\,|$ by extending the ones given in Lemma's 4.1 and 4.2 with

$$|qs(xs, ys)| = |xs|.$$

Next, we define a Herbrand interpretation of `QS` by putting

$$
\begin{aligned}
I \ = \ & \{qs(xs, ys)|\ |xs| = |ys|\} \\
& \cup\ \{f(x, xs, y1s, y2s)\ |\ |xs| = |y1s| + |y2s|\} \\
& \cup\ \{a(xs, ys, zs)\ |\ |xs| + |ys| = |zs|\} \\
& \cup\ [X > Y] \\
& \cup\ [X \leq Y].
\end{aligned}
$$

Recall that $[A]$ for an atom A stands for the set of all ground instances A.

Lemma 4.4 I *is a model of* `QS`.

Proof. First, note that $|[]| + |ys| = |ys|$ and that $|xs| + |ys| = |zs|$ implies $|[x|xs]| + |ys| = |[x|zs]|$. This implies that I is a model of `append`.

Next, note that $|[\,]| + |[\,]| = |[\,]|$ and that $|xs| = |y1s| + |y2s|$ implies $|[y|xs]| = |[y|y1s]| + |y2s|$ and $|[y|xs]| = |y1s| + |[y|y2s]|$. This implies that I is a model of `filter`.

Finally, note that $|[]| = |[]|$ and that $|xs| = |x1s| + |x2s|$, $|x1s| = |y1s|$, $|x2s| = |y2s|$ and $|y1s| + |[x|y2s]| = |ys|$ imply $|[x|xs]| = |ys|$. This implies that I is a model of `QS`. □

We now prove the desired result.

Theorem 4.5 `QS` *is acceptable w.r.t.* $|\,|$ *and* I.

Proof. As `filter` and `append` are recurrent w.r.t. $|\,|$, we only need to consider clauses (qs_1) and (qs_2). (qs_1) satisfies the appropriate requirement voidly.

Consider now a ground instance C of (qs_2). C is of the form $A \leftarrow B_1, B_2, B_3, B_4$. We now prove three facts which obviously imply that C satisfies the appropriate requirement.

Fact 1 $|A| > |B_1|$.

Proof. Note that

$$|qs([x|xs], ys)]| = |[x|xs]| > |xs| = |f(x, xs, x1s, x2s)|.$$

$\square$

Fact 2 Suppose $I \models B_1$. Then $|A| > |B_2|$ and $|A| > |B_3|$.

Proof. By assumption $|xs| = |x1s| + |x2s|$, so

$$|qs([x|xs], ys)| > |xs| \geq |x1s| = |qs(x1s, y1s)|$$

and analogously

$$|qs([x|xs], ys)| > |qs(x2s, y2s)|.$$

$\square$

Fact 3 Suppose $I \models B_1$ and $I \models B_2$. Then $|A| > |B_4|$.

Proof. By Fact 2 $|qs([x|xs], ys)| > |qs(x1s, y1s)| = |x1s|$ and by assumption $|x1s| = |y1s|$, so

$$|qs([x|xs], ys)| > |y1s| = |a(y1s, [x|y2s], ys)|.$$

$\square$

$\square$

So far we only proved that QS is left terminating. We now prove that it terminates for a large class of goals.

Lemma 4.6 *For all terms $t, t_1, \ldots, t_k$, $k \geq 0$, a goal of the form*

$$\leftarrow qs([t_1, \ldots, t_k], t)$$

is bounded w.r.t. $| \; |$ *and* I.

Proof. Let A be a ground instance of $qs([t_1, \ldots, t_k], t)$. Then $|A| = |[t_1, \ldots, t_k]| = k$, so $| \leftarrow A|_I = \text{bag}(k)$. Hence $\leftarrow qs([t_1, \ldots, t_k], t)$ is bounded by k w.r.t. $| \; |$ and I. $\square$

It is worth noting that every "ill typed" goal $\leftarrow qs(s, t)$, where s is a non-variable, non-list term is also bounded w.r.t. $| \; |$ and I, as $|s'| = 0$ for every ground instance s' of s.

Corollary 4.7 *For all terms $t, t_1, \ldots, t_k$, $k \geq 0$, all LD-derivations of* QS $\cup \{\leftarrow qs([t_1, \ldots, t_k], t)\}$ *are finite.*

Proof. By Corollary 3.8. $\square$

Permutation

Consider now the following program PERM (for permutation) studied in Plümer [Plü90b] :

```
(p₁)  p([], [])←.
(p₂)  p(Xs, [X | Ys]) ←
          a(X1s, [X|X2s], Xs),
          a(X1s, X2s, Zs),
          p(Zs, Ys)).
```

augmented by the clauses (a_1) and (a_2) which form the **append** program defining the relation a.

The intention is to invoke p with its first argument instantiated. Clause (p_1) states that the empty list is a permutation of itself. Clause (p_2) takes care of a non-empty list xs - one should first split it into two sublists $x1s$ and $[x|x2s]$ and concatenate $x1s$ and $x2s$ to get zs. If now ys is a permutation of zs, $[x|ys]$ is a permutation of xs.

Lemma 4.8 PERM *is not recurrent.*

Proof. By Theorem 2.8 of Bezem [Bez89] every recurrent program P is terminating, which means that all SLD-derivations of P starting with a ground goal are finite. But the SLD-derivation of PERM $\cup\{\leftarrow p(xs, [x|ys])\}$ with xs, x, ys ground, in whose second goal the middle atom $a(x1s, x2s, zs)$ is selected, diverges when clause (a_2) is repeatedly used. Thus PERM is not terminating and so it is not recurrent. □

We now prove that PERM is acceptable. First, we define a level mapping by putting

$$|p(zs, ys)| = |zs| + 1,$$
$$|a(x1s, x2s, zs)| = \min(|x1s|, |zs|).$$

Next, we define a Herbrand interpretation I by putting

$$I = \quad [p(Zs, Ys)]$$
$$\cup \ \{a(x1s, x2s, zs) \,|\, |x1s| + |x2s| = |zs|\}.$$

Lemma 4.9 I *is a model of* PERM.

Proof. I is trivially a model of (p_1) and (p_2). In the proof of Lemma 4.4 we showed that I is also a model of **append**. □

We can now prove the desired result.

Theorem 4.10 PERM *is acceptable w.r.t.* $|\ |$ *and* I.

Proof. It is easy to see that **append** is recurrent w.r.t. $|\ |$, so we only need to consider clause (p_2). Let $C = A \leftarrow B_1, B_2, B_3$ be a ground instance of (p_2). The required condition for C is implied by the following three facts.

Fact 1 $|A| > |B_1|$.

Proof. Note that

$$|p(xs, [x|ys])| = |xs| + 1 > |xs| \geq \min(|x1s|, |xs|) = |a(x1s, [x|x2s], xs)|.$$

$\square$

Fact 2 Suppose $I \models B_1$. Then $|A| > |B_2|$.

Proof. By assumption $|x1s| + |[x|x2s]| = |xs|$, so

$$|p(xs, [x|ys])| = |xs| + 1 > |x1s| \geq \min(|x1s|, |zs|) = |a(x1s, x2s, zs)|.$$

$\square$

Fact 3 Suppose $I \models B_1$ and $I \models B_2$. Then $|A| > |B_3|$.

Proof. By assumption $|x1s| + |[x|x2s]| = |xs|$ and $|x1s| + |x2s| = |zs|$, so

$$|p(xs, [x|ys])| = |xs| + 1 > |xs| = |x1s| + |x2s| + 1 = |zs| + 1 = |p(zs, ys)|.$$

$\square$

$\square$

Also, we have the following.

Lemma 4.11 *For all terms* $t, t_1, \ldots, t_k$, $k \geq 0$, *a goal of the form*

$$\leftarrow p([t_1, \ldots, t_k], t)$$

is bounded w.r.t. $| \ |$ *and* I.

Proof. The same as that of Lemma 4.6. $\square$

Corollary 4.12 *For all terms* $t, t_1, \ldots, t_k$, $k \geq 0$, *all LD-derivations of* PERM $\cup \{\leftarrow p([t_1, \ldots, t_k], t)\}$ *are finite.*

Proof. By Corollary 3.8. $\square$

It is useful to note that we had to use here for append a different level mapping than the one used in the proof of acceptability of QS. With the original level mapping for append, PERM is not acceptable w.r.t. any model. Indeed, consider a ground instance A of the head of (p_2). Let $C = A \leftarrow B_1, B_2, B_3$ be a ground instance of (p_2) in which the variable $X1s$ is instantiated to some ground term t with $|t| = |A|$. Then with the original level mapping for append we have $|A| = |t| = |B_1|$.

In contrast, the level mapping for append used in Theorem 4.10 can also be used in the proof of acceptability of QS.

Mergesort

Finally, consider the following program MS (for mergesort) taken from Ullman and Van Gelder [UvG88]:

(ms_1) `ms([], []) ←.`
(ms_2) `ms([X], [X]) ←.`
(ms_3) `ms([X | [Y | Xs]], Ys) ←`
          `s([X | [Y | Xs]], X1s, X2s),`
          `ms(X1s, Y1s),`
          `ms(X2s, Y2s),`
          `m(Y1s, Y2s, Ys).`

(s_1) `s([], [], []) ←.`
(s_2) `s([X | Xs], [X | Ys], Zs) ←`
          `s(Xs, Zs, Ys).`

(m_1) `m([], Xs, Xs) ←.`
(m_2) `m(Xs, [], Xs) ←.`
(m_3) `m([X | Xs], [Y | Ys], [X | Zs]) ←`
          `X ≤ Y,`
          `m(Xs,[Y | Ys], Zs).`
(m_4) `m([X | Xs], [Y | Ys], [Y | Zs]) ←`
          `X > Y,`
          `m([X | Xs], Ys, Zs).`

We assume that MS operates on the same domain as QS. The intention is to invoke ms with its first argument being an unsorted list. Clause (ms_3) takes care of non-empty list of length at least 2. The idea is first to split the input list in two lists of roughly equal length (note the reversed order of parameters in the recursive call of s), then mergesort each sublist and finally merge the resulting sorted sublists.

Denote the program consisting of the clauses $(s_1), (s_2)$ by split, and the program consisting of the clauses $(m_1), (m_2), (m_3), (m_4)$ by merge.

Lemma 4.13 split *is recurrent with* $|s(xs, x1s, x2s)| = |xs|$. ☐

Lemma 4.14 merge *is recurrent with* $|m(xs, ys, zs)| = |xs| + |ys|$. ☐

Lemma 4.15 MS *is not recurrent.*

Proof. Analogous to that of Lemma 4.3. ☐

We now show that MS is acceptable. We define an appropriate level mapping $|\ |$ by extending the ones given in Lemma's 4.13 and 4.14 with

$$|ms(xs, ys)| = |xs| + 1.$$

Next, we define a Herbrand interpretation of MS by putting

$$
\begin{aligned}
I \quad = \quad & \{ms(xs, ys)\,|\ |xs| = |ys|\} \\
& \cup\ \{s(xs, y1s, y2s)\ |\ |y1s| = \lceil|xs|/2\rceil, |y2s| = \lfloor|xs|/2\rfloor\} \\
& \cup\ \{m(xs, ys, zs)\ |\ |xs| + |ys| = |zs|\} \\
& \cup\ [X > Y] \\
& \cup\ [X \leq Y].
\end{aligned}
$$

Lemma 4.16 *I is a model of* MS.

Proof. First, note that $\|[\,]\| + |xs| = |xs|$, $|xs| + \|[\,]\| = |xs|$, $|xs| + \|[y|ys]\| = |zs|$ implies $\|[x|xs]\| + \|[y|ys]\| = \|[x|zs]\|$, and that $\|[x|xs]\| + |ys| = |zs|$ implies $\|[x|xs]\| + \|[y|ys]\| = \|[y|zs]\|$. This implies that I is a model of merge.

Next, note that $\|[\,]\| = \|[\,]\|$ and $\|[x]\| = \|[x]\|$ imply that I is a model of (ms_1) and (ms_2). Moreover, $|x1s| = \lceil|[x|[y|xs]]|/2\rceil$ and $|x2s| = \lfloor|[x|[y|xs]]|/2\rfloor$ imply $\|[x|[y|xs]]\| = |x1s| + |x2s|$, which, together with $|x1s| = |y1s|, |x2s| = |y1s|$ and $|y1s| + |y2s| = |ys|$, imply $\|[x|[y|xs]]\| = |ys|$. This implies that I is a model of (ms_3).

Next, note that $\|[\,]\| = \lceil|[\,]|/2\rceil$ and $\|[\,]\| = \lfloor|[\,]|/2\rfloor$ imply that I is a model of (s_1). Finally, to see that I is a model of (s_2), consider an atom $s(xs, zs, ys) \in I$. The following two cases arise.

Case 1 $|xs| = 2k, k \geq 0$. By assumption, $|zs| = k$ and $|ys| = k$. This implies $\|[x|ys]\| = k + 1 = \lceil|[x|xs]|/2\rceil$ and $|zs| = k = \lfloor|[x|xs]|/2\rfloor$.

Case 2 $|xs| = 2k + 1, k \geq 0$. By assumption, $|zs| = k + 1$ and $|ys| = k$. This implies $\|[x|ys]\| = k + 1 = \lceil|[x|xs]|/2\rceil$ and $|zs| = k + 1 = \lfloor|[x|xs]|/2\rfloor$.

In both cases we conclude that $s([x|xs], [x|ys], zs) \in I$, i.e. I is a model of (s_2). $\qquad\square$

We now prove the desired result.

Theorem 4.17 MS *is acceptable w.r.t.* $|\,|$ *and* I.

Proof. As split and merge are recurrent w.r.t. $|\,|$, we only need to consider clauses $(ms_1), (ms_2)$ and (ms_3). (ms_1) and (ms_2) satisfy the appropriate requirement voidly.

Consider now a ground instance $C = \leftarrow B_1, B_2, B_3, B_4$ of (ms_3). We prove three facts which imply that C satisfies the appropriate requirement.

Fact 1 $|A| > |B_1|$.

Proof. Note that

$$
|ms([x|[y|xs]], ys)])| = \|[x|[y|xs]]\| + 1 > \|[x|[y|xs]]\| = |s([x|[y|xs]], x1s, x2s)|.
$$

$\qquad\square$

Fact 2 Suppose $I \models B_1$. Then $|A| > |B_2|$ and $|A| > |B_3|$.

Proof. By assumption $|x1s| = \lceil|[x|[y|xs]]|/2\rceil$ and $|x2s| = \lfloor|[x|[y|xs]]|/2\rfloor$, which implies $\|[x|[y|xs]]\| > |x1s|$ and $\|[x|[y|xs]]\| > |x2s|$, as $\|[x|[y|xs]]\| > 1$. Hence

$$
|ms([x|[y|xs]], ys)| = \|[x|[y|xs]]\| + 1 > |x1s| + 1 = |ms(x1s, y1s)|
$$

and analogously

$$|ms([x|[y|xs]], ys)| > |ms(x2s, y2s)|.$$

$\square$

Fact 3 Suppose $I \models B_1$, $I \models B_2$ and $I \models B_3$. Then $|A| > |B_4|$.

Proof. By assumption $|ms([x|[y|xs]], ys)| > |[x|[y|xs]]| = |x1s| + |x2s|$ and $|x1s| = |y1s|$, $|x2s| = |y2s|$, so

$$|ms([x|[y|xs]], ys)| > |y1s| + |y2s| = |m(y1s, y2s, ys)|.$$

$\square$

$\square$

Additionally, we have the following.

Lemma 4.18 *For all terms $t, t_1, \ldots, t_k$, $k \geq 0$, a goal of the form*

$$\leftarrow ms([t_1, \ldots, t_k], t)$$

is bounded w.r.t. $|\ |$ and I.

Proof. The same as that of Lemma 4.6.

$\square$

Corollary 4.19 *For all terms $t, t_1, \ldots, t_k$, $k \geq 0$, all LD-derivations of* MS $\cup \{\leftarrow ms([t_1, \ldots, t_k], t)\}$ *are finite.*

Proof. By Corollary 3.8.

$\square$

5 Conclusions

Assessment of the method

Our approach to termination is limited to the study of left terminating programs, so it is useful to reflect how general this class of programs is. The main result of Bezem [Bez89] states that every total recursive function can be computed by a recurrent program. As recurrent programs are left terminating, the same property is shared by left terminating programs.

For a further analysis of left terminating programs we first introduce the following notions, essentially due to Dembinski and Maluszynski [DM85]. We follow here the presentation of Plümer [Plü90a]. Given an n-ary relation symbol p, by a *mode* for p we mean a function d_p from $\{1, \ldots, n\}$ to the set $\{+, -\}$. We write d_p in a more suggestive form $p(d_p(1), \ldots, d_p(n))$.

Modes indicate how the arguments of a relation should be used. If $d_p(i) = \text{`}+\text{'}$, we call $i$ the *input position* of $p$ and if $d_p(i) = \text{`}-\text{'}$, we call i the *output position* of p (both w.r.t. d_p) . The input positions should be replaced by ground terms and the output positions by variables. This motivates the following notion.

Given a mode d_p for a relation p, we say that an atom $A = p(t_1, \ldots, t_n)$ *respects* d_p if for $i \in [1, n]$, t_i is ground if i is an input position of p w.r.t. d_p and t_i is a variable if i is an output position of p w.r.t. d_p.

A *mode* for a program P is a function which assigns to each relation symbol of P a non-empty set of modes. Given a mode for a program P, we say that an atom A *respects moding* if A respects some mode in the set of modes associated with the relation p used in A.

As an example consider the mode for the program append represented by the following set:

$$\{\text{append}(+, +, -), \quad \text{append}(-, -, +)\}.$$

It indicates that append should be called either with its first two arguments ground and the third being a variable, or with its first two arguments being a variable and the third argument ground. Then any atom $append(xs, ys, zs)$, where either xs, ys are ground and zs is a variable, or xs, ys are variables and zs is ground, respects moding.

The following simple theorem shows that the property of left termination is quite natural.

Theorem 5.1 *Let P be a program with a mode such that for all atoms A which respect moding, all LD-derivations of $P \cup \{\leftarrow A\}$ are finite. Then P is left terminating.*

Proof. Consider a ground atom A. A is a ground instance of some atom B which respects moding. By a variant of the Lifting Lemma applied to the LD-resolution we conclude that all LD-derivations of $P \cup \{\leftarrow A\}$ are finite. This implies that P is left terminating. $\square$

The assumptions of the above theorem are satisfied by an overwhelming class of Prolog programs.

As Theorem 3.11 shows, the method presented in this paper is a complete method for proving termination of Prolog programs. We believe that it is also a useful method, since it allows us to factore termination proofs into simpler, separate proofs, which consist of checking the guesses for the level mapping $| \ |$ and the model I. Moreover, the method is modular, because termination proofs provided for subprograms can be reused in later proofs.

In this paper, the method is used as an "a posteriori" technique for verifying termination of existing Prolog programs. However, it could also provide a guideline for the program development, if the program is constructed together with its termination proof. A specific level mapping and a model could suggest, in particular, a specific ordering of atoms in clause bodies.

It is worth noting that some fragments of the proof of accceptability can be automated, at least in the case of the applications presented in Section 4. In our examples, where the function *listsize* is used, the task of checking the guesses for both the level mapping $| \ |$ and the model I can be reduced to checking the validity of universal formulas in Presburger arithmetic, which is a decidable theory. To illustrate this point, consider the following guess I for a model for the program PERM:

$$I = \quad \{p(zs, ys) \mid |zs| = |ys|\}$$
$$\cup \ \{a(x1s, x2s, zs) \mid |x1s| + |x2s| = |zs|\}.$$

To show that I is a model of, say, clause (p_2), we have to prove the following implication:

$$\{a(x1s, [x|x2s], xs), a(x1s, x2s, zs), p(zs, ys)\} \subseteq I \Rightarrow p(xs, [x|ys]) \in I.$$

By homomorphically mapping lists onto their lengths, i.e. by mapping $[\,]$ to 0 and $[: \mid .]$ to the successor function $s(.)$, we get the following formula of Presburger arithmetic:

$$n_1 + n_2 + 1 = n \ \wedge \ n_1 + n_2 = k \ \wedge \ k = m \Rightarrow n = m + 1$$

where $n_1 = |x1s|, n_2 = |x2s|, n = |xs|, k = |zs|, m = |ys|$.

Analogous considerations apply to the verification of the level mapping.

Finally, it is useful to notice a simple consequence of our approach to termination. By proving that a program P is acceptable and a goal G is bounded, we can conclude by Corollary 3.8 that the LD-tree for $P \cup \{G\}$ is finite. Thus, for the leftmost selection rule, the set of computed answer substitutions for $P \cup \{G\}$ is finite and consequently, by virtue of the strong completeness of SLD-resolution, we can use the LD-resolution to compute the set of all correct answer substitutions for $P \cup \{G\}$. In other words, query evaluation of bounded goals can be implemented using pure Prolog.

Related work

Of course the subject of termination of Prolog programs has been studied by others. Without aiming at completeness we mention here the following related work.

Vasak and Potter [VP86] identified two forms of termination for logic programs – existential and universal one and characterized the class of universal terminating goals for a given program with selected selection rules. However, this characterization cannot be easily used to prove termination. Using our terminology, given a program P, a goal G is universally terminating w.r.t. a selection rule R if the SLD-tree for $P \cup \{G\}$ via R is finite.

Baudinet [Bau88] presented a method for proving termination of Prolog program in which with each program a system of equations is associated whose least fixpoint is the meaning of the program. By analyzing this least fixpoint various termination properties can be proved. The main method of reasoning is fixpoint or structural induction.

Ullman and Van Gelder [UvG88] considered the problem of automatic verification of termination of a Prolog program and a goal. In their approach first some sufficient set of inequalities between the sizes of the arguments of the relation symbols are generated, and then it is verified if they indeed hold. Termination of the programs studied in the previous section is beyond the scope of their method.

This approach was improved in Plümer [Plü90b], [Plü90a], who allowed a more general form of the inequalities and the way sizes of the arguments are measured. This resulted in a more powerful method. Both the quicksort and the permutation programs studied in the previous section can be handled using Plümer's method. However, the mergesort remains beyond its scope.

Deville [Dev90] also considers termination in his proposal of systematic program development. In his framework, termination proofs exploit well-founded orderings together with mode and multiplicity informa-tion, the latter representing an upper bound to the number of answer substitutions for goals which respect a given mode. For instance, a termination proof of the program DC of Example 3.1(iii) for the goal $\leftarrow dc(x, Y)$ would involve verification of the following statements (assuming that x is a ground term):

1. the goal $\leftarrow divide(x, X1, X2)$ respects moding, and both $X1$ and $X2$ are bound to ground terms, $x1$ and $x2$ respectively, by any computed answer substitution for such a goal;

2. both $x1$ and $x2$ are smaller than x w.r.t. some well-founded ordering;

3. the mode $divide(+, -, -)$ has a finite multiplicity.

Our approach seems to be simpler as it relies on fewer concepts. Also, it suggests a more uniform methodology. On the other hand, in Deville's approach more information about the program is obtained.

References

[Apt88] K. R. Apt. Introduction to logic programming. Technical Report CS-R8826, Centre for Mathematics and Computer Science, 1988. To appear in "Handbook of Theoretical Computer Science", North Holland (J. van Leeuwen, ed.).

[AvE82] K. R. Apt and M. H. van Emden. Contributions to the theory of logic programming. *J. ACM*, 29(3):841–862, 1982.

[Bau88] M. Baudinet. Proving termination properties of PROLOG programs. In *Proceedings of the 3rd Annual Symposium on Logic in Computer Science (LICS)*, pages 336–347, Edinburgh, Scotland, 1988.

[Bez89] M. Bezem. Characterizing termination of logic programs with level mappings. In E. L. Lusk and R. A. Overbeek, editors, *Proceedings of the North American Conference on Logic Programming*, pages 69–80. The MIT Press, 1989.

[Bla86] H. A. Blair. Decidability in the Herbrand Base. Manuscript (presented at the Workshop on Foundations of Deductive Databases and Logic Programming, Washington D.C., August 1986), 1986.

[Cav89] L. Cavedon. Continuity, consistency, and completeness properties for logic programs. In G. Levi and M. Martelli, editors, *Proceedings of the Sixth International Conference on Logic Programming*, pages 571–584. The MIT Press, 1989.

[Dev90]   Y. Deville. *Logic Programming. Systematic Program Development*. International Series in Logic Programming. Addison-Wesley, 1990.

[DM85]   P. Dembinski and J. Maluszynski. AND-parallelism with intelligent backtracking for annotated logic programs. In *Proceedings of the International Symposium on Logic Programming*, pages 29–38, Boston, 1985.

[Fit85]   M. Fitting. A Kripke-Kleene semantics for general logic programs. *Journal of Logic Programming*, 2:295–312, 1985.

[Flo67]   R. W. Floyd. Assigning meanings to programs. In *Proceedings Symposium on Applied Mathematics, 19, Math. Aspects in Computer Science*, pages 19–32. American Society, 1967.

[Kle52]   S. C. Kleene. *Introduction to Metamathematics*. van Nostrand, New York, 1952.

[Llo87]   J. W. Lloyd. *Foundations of Logic Programming*. Springer-Verlag, Berlin, second edition, 1987.

[Plü90a]   L. Plümer. *Termination Proofs for Logic Programs*. Lecture Notes in Artificial Intelligence 446, Springer-Verlag, Berlin, 1990.

[Plü90b]   L. Plümer. Termination proofs for logic programs based on predicate inequalities. In D. H. D. Warren and P. Szeredi, editors, *Proceedings of the Seventh International Conference on Logic Programming*, pages 634–648. The MIT Press, 1990.

[UvG88]   J. D. Ullman and A. van Gelder. Efficient tests for top-down termination of logical rules. *J. ACM*, 35(2):345–373, 1988.

[VP86]   T. Vasak and J. Potter. Characterization of terminating logic programs. In *Proceedings of the 1986 IEEE Symposium on Logic Programming*, 1986.

Concept Logics

Franz Baader[1], Hans-Jürgen Bürckert[1], Bernhard Hollunder[1],
Werner Nutt[1], Jörg H. Siekmann[2]

Abstract: Concept languages (as used in BACK, KL-ONE, KRYPTON, LOOM) are employed as knowledge representation formalisms in Artificial Intelligence. Their main purpose is to represent the generic concepts and the taxonomical hierarchies of the domain to be modeled. This paper addresses the combination of the fast taxonomical reasoning algorithms (e.g. subsumption, the classifier etc.) that come with these languages and reasoning in first order predicate logic. The interface between these two different modes of reasoning is accomplished by a new rule of inference, called *constrained resolution*. Correctness, completeness as well as the decidability of the constraints (in a restricted constraint language) are shown.

Keywords: concept description languages, KL-ONE, constrained resolution taxonomical reasoning, knowledge representation languages.

1   Introduction

All knowledge representation (KR) languages in Artificial Intelligence (AI) have one deficiency in common: None of them is adequate to represent every kind of knowledge, but instead the systems that support a particular representation formalism usually specialize on some mode of reasoning, while leaving others out.

---

[1]DFKI Kaiserslautern, Projektgruppe WINO, Postfach 2080, D-6750 Kaiserslautern, FR Germany, E-mail: {baader, buerkert, hollunde, nutt}@informatik.uni-kl.de
[2]Universität Kaiserslautern, FB Informatik, Postfach 3049, D-6750 Kaiserslautern, FR Germany, E-mail: siekmann@informatik.uni-kl.de

This is in particular the case for those knowledge representation languages that are built on some extension of first order logic on the one hand, and the class of concept languages of the KL-ONE family on the other hand. The systems that support either of these two classes of formalisms have highly tuned and specialized inference mechanisms that pertain to the task at hand, but are utterly inefficient (or outright impossible) when it comes to reasoning in a mode that is more typical for the other class. Hence one would like a combination of the KR-formalisms for a more satisfactory representation of the knowledge domain.

What is required are hybrid KR-formalisms that support an adequate representation of different kinds of knowledge chunks. One step in this direction has been made in the field of automated deduction, where sorted logics have been proposed, i.e., an extension of standard predicate logic (PL1) with so-called sort hierarchies (Frisch, 1989; Walther, 1987; Schmidt-Schauß, 1989; Beierle et al., 1990; Weidenbach & Ohlbach, 1990; Cohn, 1987). Sort hierarchies are set description languages with a relatively weak expressiveness[3], but they turned out to provide a powerful extension of predicate logic that can still be based on the Resolution Principle (Robinson, 1965). It has been shown that sort hierarchies – or their operationalization via sort resolution – allow a substantial pruning of the search space by replacing certain deduction sequences with a deterministic computation of the subset relationships (Frisch, 1986; Walther, 1988). The essential idea is this: based on work carried out in the logic community in the 1930's and 40's (Herbrand, 1930; Oberschelp, 1962; Schmidt, 1938, 1951) the universe of discourse (the domain for the interpretation) is partitioned into several classes (the sorts) in the case of many sorted logics, which again may have subsorts in the case of order sorted logics. Instead of a unary predicate that expresses this information, sort symbols are attached to variables, constants etc. For example the fact that all natural numbers are either even or odd can be expressed as:

$$\forall x.\ NAT(x)\ \text{implies}\ EVEN(x)\ \text{or}\ ODD(x)$$

or in a sorted logic as

$$\forall x{:}NAT\ .\ EVEN(x)\ \text{or}\ ODD(x).$$

Since there is a general translation (called relativization) between these variants of logic and since also the expressive power is not principally enhanced but remains that of a first order language, interest in the logic community faded by the mid sixties.

This work was rediscovered however by workers in the field of automated reasoning (Walther 1987), when it became apparent that although the expressive power does not increase in principle, there is still an enormous difference in practice, when it comes to actually building an inference engine for the respective logics.

The same general observation holds for current concept description languages in AI and their relationship to predicate logic: while they do not enhance the expressiveness of PL1 in

---

[3] Sorts are interpreted as sets, sort hierarchies specify the subset relationship.

principle, they definitely advance the state of the art, when it comes to actually reason – i.e. draw conclusions – in their respective formalisms.

Concept description formalisms of the KL-ONE family (Brachman & Schmolze, 1985; Brachman & Levesque, 1985; Vilain, 1985; MacGregor & Bates, 1987; Nebel, 1990) are more expressive than mere sort hierarchies. They provide terminologies of concepts or sets whose underlying taxonomical hierarchies determine subset relationships quite similar to sort hierarchies. But in contrast to sort hierarchies concept languages allow far more complex constructs for set descriptions themselves. We shall show one possible combination of this terminological reasoning with that of standard predicate logic, based on the idea to enhance PL1 with concept descriptions, which are then taken as constraints for the variables occurring in the formulae – just as in the above mentioned approaches to sorted logics. However, there are two main differences: The first is that these constraints have a more complex structure: sets are not just denoted by sort symbols as in sorted logics, but they are denoted by complex *concept description terms*. The second difference is that instead of simple sort hierarchies we now have more complicated taxonomies, which do not necessarily possess generic models (least Herbrand models) as in the case of ordinary sort hierarchies (Frisch, 1989).

This paper addresses the combination of the fast taxonomical reasoning algorithms (e.g. subsumption, the classifier etc.) that come with these languages and reasoning in first order predicate logic. The interface between these two different modes of reasoning is accomplished by a constrained-based modification of the Resolution Principle. This can be seen as a generalization of M. Stickel's theory resolution principle (Stickel, 1985) on the one hand and of the ideas of constrained logic programming (Höhfeld & Smolka, 1988; Jaffar & Lassez, 1987) on the other hand, and is worked out in (Bürckert, 1990).

2 Constrained Resolution

The resolution principle elaborates the idea that we can infer the *resolvent* formula $B \vee C$, from the *parent* formulae $A \vee B$ and $\neg A \vee C$. Here the formulae have to be clauses, i.e., universally quantified disjunctions of literals. The A's of the two parent formulae are complementary literals, and in order to obtain the resolvent we have to *unify* corresponding arguments of those literals and apply the *unifying substitution* to the resolvent, for example:

$$\frac{P(s_1,\ldots,s_n) \vee B \qquad \neg P(t_1,\ldots,t_n) \vee C}{\sigma(B \vee C)} \quad \text{if there is some } \sigma \text{ with } \sigma s_i = \sigma t_i \ (1 \leq i \leq n)^4$$

---

[4] This inference rule has to be read as follows: From the formulae above the line infer the formulae below the line provided the condition can be satisfied.

An analysis of Robinson's soundness and completeness proof for resolution (Robinson, 1965), provides an argument that the computation of a unifying substitution can be replaced by a unifiability test, provided we add a *constraint* consisting of the conjunction of the term equations $s_i = t_i\ (1 \leq i \leq n)$ to the resolvent (instead of instantiating it with the substitution)[5].

A slight modification of this approach transformes the clauses into homogeneous form, where the argument terms of the literals are replaced by new variables. Then constraints are attached to the clauses, that specify the equality of these new variables with the substituted terms. For example

$$P(b, x) \lor Q(a, f(x, b), g(z))$$

is replaced by

$$P(x_1, x_2) \lor Q(x_3, x_4, x_5)\ //\ x_1=b,\ x_2=x,\ x_3=a,\ x_4=f(x, b),\ x_5=g(z).$$

Now a resolution step takes two clauses with such *equational constraints* and generates a resolvent with a new unified equational constraint.

Noticing that unifiability of equations is the same as the satisfiability of the existential closure of these equations in the (ground) term algebra, we obtain a more general view: clauses might have some arbitrary, not necessarily equational constraint for their variables. A resolvent of two clauses generates a new constraint that is unsatisfiable whenever one of the parents' constraints is unsatisfiable. Here the derived new constraint is a conjunction of the old ones after an identification of the corresponding variables. Let Γ and Γ' denote constraints that are separated from their corresponding clause by $//$, then we denote constrained resolution as:

$$\frac{P(x_1,\ldots,x_n) \lor C\ //\ \Gamma \qquad \neg P(y_1,\ldots,y_n) \lor C'\ //\ \Gamma'}{C \lor C'\ //\ \Gamma \land \Gamma'_{[y_i\ =\ x_i\ (1 \leq i \leq n)]}} \quad \text{if } \Gamma \land \Gamma'_{[y_i\ =\ x_i\ (1 \leq i \leq n)]} \text{ is satisfiable}$$

The satisfiability of the constraint is defined with respect to a distinct constraint theory.

As in the classical case, the overall task is now to prove the unsatisfiability of a given set of constrained clauses, where we have to take into account that the constraints are interpreted with respect to the given constraint theory. This constraint theory could be any set of models – either given by a set of axioms or in any other way. In the classical case for instance, the constraint theory might be thought of as a singleton set containing just the ground term algebra, i.e., the Herbrand universe.

Let us introduce these notions of constraint theories and constrained clauses more formally. A *constraint theory* consists of a set of *constraint symbols* (or constraints) and a set of *constraint models*, in which these constraint symbols are interpreted (cf. also Höhfeld & Smolka, 1988; Smolka, 1989). Given an infinite set $\mathcal{V}$ of variables we assume that

---

[5] Huet (Huet, 1972) in fact used this trick in order to have a resolution rule for higher order logics. He already used the name "constrained resolution" for this rule.

every constraint Γ comes with a finite set of variables $Var(\Gamma) = \{x_1,...,x_n\}$. In this case the constraint is often denoted by $\Gamma(x_1,...,x_n)$. Each model $\mathcal{A}$ consists of a nonempty set $D^{\mathcal{A}}$, the domain, and for each constraint symbol Γ it contains a set $\Gamma^{\mathcal{A}}$ of $\mathcal{A}$-*solutions*, i.e., of assignments $\sigma: \mathcal{V} \to D^{\mathcal{A}}$. Up to now a constraint theory is nothing but a set of first order structures over a (possibly infinite) signature of predicate symbols, namely the constraint symbols. The set of n-tuples, given by the $\mathcal{A}$-solution values of the variables of an n-ary constraint, is just the denotation of the corresponding predicate symbol in the first order structure $\mathcal{A}$. However, for the constrained resolution rule we need in addition that the set of constraints has to be closed under conjunction, under renaming[6] and under identification of variables. That is, for every two constraints there must be a conjunct constraint, whose solution sets are the intersections of its constituents. Also for each constraint and each variable substitution, which renames or identifies variables, there must be a variant or an instance constraint, whose solutions are obtained by renaming the former solutions accordingly or by selecting just those solutions that assign the same value to identified variables. To be more precise: we assume that for every two constraints Γ_1 and Γ_2 there exists their conjunct constraint $\Gamma_1 \wedge \Gamma_2$ such that $(\Gamma_1 \wedge \Gamma_2)^{\mathcal{A}} = \Gamma_1^{\mathcal{A}} \cap \Gamma_2^{\mathcal{A}}$ for every constraint model $\mathcal{A}$; and we assume that for every constraint Γ and for every (renaming or identifying) variable substitution $\varphi: \mathcal{V} \to \mathcal{V}$ there exists the (variant or instance) constraint $\varphi\Gamma$ such that $(\varphi\Gamma)^{\mathcal{A}} = \{\sigma^* \mid \sigma \in \Gamma^{\mathcal{A}}$ with $\sigma x = \sigma y$ if $\varphi x = \varphi y\}$ for every constraint model $\mathcal{A}$, where σ^* is defined by $\sigma^*\varphi x := \sigma x$ for each variable $\varphi x \in \varphi(\mathcal{V})$, and $\sigma^* z := \sigma z$ for all other variables z.

We now want to extend constraint theories with further predicate symbols. Thus given a constraint theory and a set $\mathcal{P}$ of predicate symbols – disjoint from the set of constraint symbols – we augment the constraint models by any possible denotation for these new symbols: An $\mathcal{R}$-*structure* is given by a constraint model, where in addition each n-ary predicate symbol $P \in \mathcal{P}$ denotes an n-ary relation $P^{\mathcal{A}}$ on $D^{\mathcal{A}}$.

For technical reasons we do not allow arbitrary formulae over these predicate symbols as there might be problems transforming them into constrained clauses, depending on the constraint theory (Bürckert, 1990). Thus we restrict ourselves to formulae that are already in (constrained) clause form, i.e., to sets of constrained clauses.

An *atom* is a predicate symbol followed by a list of variables: $P(x_1,...,x_n)$. A literal is an atom or its negation: $\neg P(x_1,...,x_n)$. A *constrained clause* is a pair consisting of a set C of literals (called the kernel or the matrix of the clause) and a constraint Γ. Constrained clauses are written $C \,//\, \Gamma$.

The semantics of constrained clauses are defined as follows.

---

[6] As in the classical case clauses have always to be variable disjoint, which requires a renaming of the generated resolvents – and hence also of their constraints – before they can be added to the clause set.

Let $\mathcal{A}$ be an $\mathcal{R}$-structure, let α be an $\mathcal{A}$-assignment, let P be a predicate symbol, and let C be a set of literals, i.e., the kernel of a constrained clause. Then

➤ $(\mathcal{A}, \alpha) \vDash P(x_1,...,x_n)$ iff $(\alpha x_1,...,\alpha x_n) \in P^{\mathcal{A}}$

➤ $(\mathcal{A}, \alpha) \vDash \neg P(x_1,...,x_n)$ iff $(\alpha x_1,...,\alpha x_n) \notin P^{\mathcal{A}}$

➤ $(\mathcal{A}, \alpha) \vDash C$ iff $(\mathcal{A}, \alpha) \vDash L$ for some literal L in the set C

Let $C \,//\, \Gamma$ be a constrained clause. Then

➤ $\mathcal{A} \vDash C \,//\, \Gamma$ iff $(\mathcal{A}, \alpha) \vDash C$ for each $\mathcal{A}$-solution α of Γ.

Note that the constrained clause is also satisfied by the $\mathcal{R}$-structure, if the constraint has no $\mathcal{A}$-solution. In particular, a constrained clause, whose constraint does not have an $\mathcal{A}$-solution in any $\mathcal{R}$-structure $\mathcal{A}$, is a tautology with respect to the constraint theory.

We call a set of constrained clauses $\mathcal{R}$-*satisfiable* iff there is an $\mathcal{R}$-structure $\mathcal{A}$ such that each of the clauses is satisfied by the structure; otherwise the clause set is called $\mathcal{R}$-*unsatisfiable*.

Not every constraint needs to have solutions, but in order to specify the derivation rules for proving the unsatisfiability of a constrained clause set our main interest lies in constraints that have solutions – at least in one constraint model. A constraint Γ is called *satisfiable* or solvable iff there is at least one model $\mathcal{A}$, where the $\mathcal{A}$-solution set of the constraint is not empty. We have the following *constrained resolution rule* for a pair of variable disjoint constrained clauses:

$$\frac{\{ P(x_{11},...,x_{1n}),...,\ P(x_{k1},...,x_{kn})\} \cup C \,//\, \Gamma \qquad \{\neg P(y_{11},...,y_{1n}),...,\neg P(y_{m1},...,y_{mn})\} \cup C' \,//\, \Gamma'}{C \cup C' \,//\, \varphi(\Gamma \wedge \Gamma')} \text{ if } \varphi(\Gamma \wedge \Gamma') \text{ is satisfiable}$$

where C and C' are the remaining parts of the two clauses and the new constraint $\varphi(\Gamma \wedge \Gamma')$ is a conjunction instantiated by a variable substitution φ that identifies corresponding variables of the complementary literals: $\varphi x_{1i} = ... = \varphi x_{ki} = \varphi y_{1i} = ... = \varphi y_{mi}$ $(1 \leq i \leq n)$. The derived clause is called a *resolvent* of the two parent clauses, its variables have to be renamed, in order to keep the clauses' variables disjoint. This whole operation on clause sets is called a *resolution step*. A *derivation* is a (possibly infinite) sequence of resolution steps.

On closer inspection we notice that the unsatisfiability of a constrained clause set cannot be proved by just deriving the empty clause, as in the classical resolution calculus. The reason is that such an empty clause might still have some constraints, which are only satisfied by *some* of the constraint models, but not by all of them. For classical resolution the derivation of the empty clause could be seen as a derivation of *false* from the starting clause set, but in the constraint resolution framework this provides just a derivation of *false* within those models that satisfy the constraints of the empty clause. The solution is to construct for each constraint model a suitable empty clause whose restriction is satisfied by that model (Bürckert, 1990). We call a set of (satisfiable) constraints *valid* iff for each constraint model $\mathcal{A}$ at least one of the constraints in the set has $\mathcal{A}$-solutions. With this notion we now define: A *refutation* is a derivation, such that the set of constraints of all derived empty clauses is valid. However, this is not an operational definition: we need terminating refutations, so we have to restrict the constraint theories to those where only a finite number of such constrained empty clauses is needed. A constraint theory is *compact* iff every valid set of constraints contains a finite subset, which is again valid. For compact theories every infinite refutation contains a finite subsequence of resolution steps that is a refutation.

The following completeness result for constrained resolution can be proved by a suitable generalization of the standard completeness proof for classical resolution (Bürckert, 1990):

Theorem: (Soundness and completeness of constrained resolution)

Let $\mathcal{R}$ be a (compact) constraint theory. A set C of constrained clauses is $\mathcal{R}$-unsatisfiable iff there exists a (finite) refutation of C.

Any theory over a given first order language – given as a set of first order structures over this language – can be seen as a constraint theory, where the open first order formulae play the role of constraints. Such a constraint is satisfiable iff its existential closure is satisfiable by some of the models of the theory. The theory is a compact constraint theory iff it has an axiomatization by a set of first order formulae (this is an easy consequence of the compactness theorem of predicate logic and the above definition of validity of constraints). Obviously a finite set of constraints is valid iff the existential closure of the disjunction of its elements is a theorem of the theory, i.e. it is satisfied by each model of the theory.

Many concept languages are known to provide a special class of theories with first order axiomatizations, such that it is decidable if a concept description denotes a nonempty set in at least one of the models of that theory. It turns out that this test is essentially the basis of the satisfiability test we need in our constrained resolution steps. We shall see that the validity of a set of concept constraints is also decidable, for a reasonable class of concept languages, to be defined below.

3   Concept Languages

Concept languages were introduced by (Brachman & Schmolze, 1985) for the formalization and structuring of semantic networks (Quillian, 1968). In particular, they are used to represent the taxonomical and conceptual knowledge of a particular problem domain. To describe this kind of knowledge, one starts with given atomic concepts and roles, and defines new concepts and roles[7] using the operations provided for by the concept language. Concepts can then be

---

[7]In this paper we do not consider the definition of complex roles. See (Baader, 1990; Hollunder & Nutt, 1990) for a discussion of role-forming constructs such as the transitive closure or intersection of roles.

considered as unary predicates which are interpreted as sets of individuals and roles as binary predicates which are interpreted as binary relations over individuals.

Examples for atomic concepts may be person and female, examples for roles may be child and friend. If logical connectives like conjunction, disjunction, and negation are present as language constructs, one can describe the concept of a man as those "persons who are not female" and represent it by the expression (person $\sqcap \neg$ female), where $\sqcap$ is the conjunction and $\neg$ is the negation operator provided by the concept language. Conjunction, disjunction, and negation are interpreted as set intersection, union, and complement. Most languages provide quantification over roles that allows for instance to describe the concepts of "individuals having a female child" and "individuals for which all children are female" by the expressions $\exists$child.female and $\forall$child.female, respectively. Number restrictions on roles denote sets of individuals having at least or at most a certain number of fillers for a role. For instance, (≥ 3 friend) $\sqcap$ (≤ 2 child) can be read as "all individuals having at least three friends and at most two children."

Concept languages, e.g., $\mathcal{FL}$ and $\mathcal{FL}^-$ (Levesque & Brachman, 1987), $\mathcal{TF}$ and $\mathcal{NTF}$ (Nebel, 1990), or the $\mathcal{AL}$-languages considered in (Schmidt-Schauß & Smolka, 1988; Hollunder & Nutt, 1990), differ in what kind of constructs are allowed for the definition of concepts and roles. Their common feature – besides the use of concepts and roles – is that the meaning of the constructs is defined by a model-theoretic semantics.

In this section we show how to conceive a concept language as a constraint theory. For this purpose we restrict our attention to the concept language $\mathcal{ALC}$ (Schmidt-Schauß & Smolka, 1988). This language offers language constructs for conjunction, disjunction, and negation of concepts as well as role quantification. It would also be possible to use an extension of $\mathcal{ALC}$, e.g. $\mathcal{ALC}$ amalgamated with number restrictions, but this would burden the presentation without any new insight.

We assume a language with two disjoint alphabets of symbols, *concept symbols* and *role symbols*. We have two special concept symbols $\top$ (*top symbol*) and $\bot$ (*bottom symbol*). The set of *concept descriptions* of $\mathcal{ALC}$ is inductively defined:

➤ every concept symbol is a concept description (*atomic description*).

Now let C and D be concept descriptions already defined and let R be a role symbol. Then

➤ $C \sqcap D$ (*conjunction*), $C \sqcup D$ (*disjunction*) and $\neg C$ (*negation*) are concept descriptions, and

➤ $\forall R.C$ (*value restriction*) and $\exists R.C$ (*exists restriction*) are concept descriptions.

An *interpretation* I of a concept description consists of a nonempty set $\top^I$ (the *domain* of I), and a function $\cdot^I$ (the *interpretation function* of I). The interpretation function maps every concept symbol A to a subset A^I of $\top^I$ and every role symbol R to a subset R^I of $\top^I \times \top^I$. The interpretation function - which gives an interpretation for concept and role symbols - will be extended to arbitrary concept descriptions as follows. Let C and D be concept descriptions and let R be a role symbol. Assume that C^I and D^I is already defined. Then

➤ $(C \sqcap D)^I \quad := \quad C^I \cap D^I$

➤ $(C \sqcup D)^I \quad := \quad C^I \cup D^I$

➤ $(\neg C)^I \quad := \quad T^I \setminus C^I$

➤ $(\forall R.C)^I \quad := \quad \{a \in T^I \mid \forall b: (a, b) \in R^I \Rightarrow b \in C^{\mathcal{A}}\}$

➤ $(\exists R.C)^I \quad := \quad \{a \in T^I \mid \exists b: (a, b) \in R^I \wedge b \in C^{\mathcal{A}}\}$.

In KL-ONE-based knowledge representation systems such as KL-ONE (Brachman & Schmolze, 1985), BACK (Nebel, 1990), KRYPTON (Brachman et al., 1985), LOOM (MacGregor & Bates, 1987), concept languages are used to describe a **terminology**, i.e., the taxonomical knowledge. Starting with concepts such as person and female one can describe "persons that are not female" by the expression (person $\sqcap$ $\neg$female). If we want to use this expression in other concept descriptions it is appropriate to define the **terminological axiom**

$$\text{man} = \text{person} \sqcap \neg\text{female}$$

where man is a concept. If child is a role, we can describe "not female persons with only female children" by the expression (man $\sqcap$ $\forall$child.female). Terminological axioms allow to define abbreviations for concept descriptions, and hence help to keep concept descriptions simple. However, for reason of simplicity of presentation we do not consider terminological axioms. Thus we assume that every concept in a concept description is completely undefined and not an abbreviation for another concept description.

A concept description C is called **satisfiable** iff there exists an interpretation I such that C^I is nonempty, and is called **universally satisfiable** iff there exists an interpretation I such that $C^I = T^I$. We say D is subsumed by C if $D^I \subseteq C^I$ for every interpretation I, and C is equivalent to D if $C^I = D^I$ for every interpretation I.

For our view of concept languages as constraint theories, we are only interested in satisfiability and universal satisfiability of concept descriptions. KL-ONE-based knowledge representation systems offer in addition to a concept language an assertional component for the introduction of objects, which are instances of concept descriptions and roles. In this case there are notions as consistency of the knowledge base, instantiation, realization and retrieval. Definitions of these and relationships between them can be found in (Nebel, 1990; Hollunder, 1990).

We shall now describe how to view a concept language as a constraint theory using concept descriptions to define socalled concept constraints. Let $\mathcal{V}$ be an infinite set of variables. An **atomic concept constraint** is of the form $x{:}C$ where x is a variable and C is a concept description. A **concept constraint** $\Gamma(x_1,\ldots,x_n)$ is a finite set $\{x_1{:}C_1,\ldots, x_n{:}C_n\}$ of atomic concept constraints.

A **solution** of a concept constraint $\Gamma = \{x_1{:}C_1,\ldots, x_n{:}C_n\}$ in an interpretation I is an I-assignment $\alpha{:}\ \mathcal{V} \to T^I$ such that $\alpha x_i \in C_i^I (1 \leq i \leq n)$. A concept constraint Γ is **satisfiable** iff there exists an interpretation I and a solution of Γ in I.

Let Γ be a concept constraint. Without loss of generality we can assume that the variables occurring in the atomic concept constraints of Γ are pairwise disjoint. To see this, suppose that $\Gamma = \{x_1{:}C_1,\ldots, x_n{:}C_n\}$ and $x_i = x_j$ for some $i \neq j$. Then we create a new concept constraint $\Gamma' := \Gamma \setminus \{x_i{:}C_i, x_j{:}C_j\} \cup \{x_i{:}(C_i \sqcap C_j)\}$. Obviously, Γ is satisfiable if and only if Γ' is satisfiable. This process is iterated until we obtain a concept constraint such that variables occurring in the atomic concept constraints are pairwise disjoint.

Now consider the constrained resolution rule: In order to make the application of the resolution step effective, we need an algorithm that decides whether the conjunction of the constraints of the parent clauses is satisfiable. In the following proposition we relate the satisfiability of concept descriptions to the satisfiability of concept constraints.

Proposition 3.1:

A concept constraint $\Gamma = \{x_1{:}C_1,\ldots, x_n{:}C_n\}$ with pairwise different variables is satisfiable if and only if every concept description C_i is satisfiable.

Proof: ($\Rightarrow$) If $\Gamma = \{x_1{:}C_1,\ldots, x_n{:}C_n\}$ is satisfiable, then there exists an interpretation I such that every C_i^I is nonempty. Hence every C_i is satisfiable.

($\Leftarrow$) Let $\Gamma = \{x_1{:}C_1,\ldots, x_n{:}C_n\}$ be a concept constraint such that $x_i \neq x_j$ for $i \neq j$. Suppose that every C_i is satisfiable. Then there exists for every C_i an interpretation I_i such that $C_i^{I_i}$ is nonempty. Without loss of generality we can assume that $T^{I_i} \cap T^{I_j} = \emptyset$ for $i \neq j$. Now, let $T^I := \bigcup_{1 \leq i \leq n} T^{I_i}, A^I := \bigcup_{1 \leq i \leq n} A^{I_i}$ for every concept symbol A, and let $R^I := \bigcup_{1 \leq i \leq n} R^{I_i}$ for every role symbol R. Then I is an interpretation and it is easy to see that Γ has a solution in I. Hence Γ is satisfiable. $\square$

The problem of checking the satisfiability of concept descriptions is considered in (Schmidt-Schauß & Smolka, 1988; Hollunder & Nutt, 1990), where it is shown that checking the satisfiability of concept descriptions of the $\mathcal{ALC}$-language is a decidable, PSPACE-complete problem.

In order to decide whether we have reached a refutation we need an algorithm that decides the validity of a finite set of concept constraints (cf. the completeness theorem of constrained resolution). We show that this can be reduced to decide the universal satisfiability of concept descriptions.

Proposition 3.2:

A set $\{\Gamma_1,\ldots,\Gamma_n\}$ of concept constraints with pairwise disjoint variables is valid iff for each sequence $(x_1{:}C_1,\ldots,x_n{:}C_n)$ with $x_i{:}C_i \in \Gamma_i (1 \leq i \leq n)$ and each interpretation I some C_i has a solution in I.

Proof: ($\Leftarrow$) We show that $\{\Gamma_1,\ldots,\Gamma_n\}$ is valid: Let I be any interpretation. Assume that none of the concept constraints is satisfiable in I. Then each Γ_i must contain an atomic constraint $x_i{:}C_i$ without a solution in I. This contradicts the precondition.

($\Rightarrow$) Assume by contradiction that there is a sequence $(x_1{:}C_1,\ldots,x_n{:}C_n)$ with $x_i{:}C_i \in \Gamma_i$ $(1 \leq i \leq n)$ and an interpretation I such that $C_i^I = \emptyset$ for each i $(1 \leq i \leq n)$. We show that $\{\Gamma_1,\ldots,\Gamma_n\}$ cannot be valid. Assume by contradiction that some Γ_{i_0} has a solution in I. Then by Proposition 3.1 all its concept descriptions are satisfiable in I, and hence especially $(C_{i_0})^I \neq \emptyset$, which yields a contradiction. $\square$

Now, let $C_1,\ldots,C_n$ be concept descriptions and let I be any interpretation. Then obviously $(\neg C_1 \sqcap \ldots \sqcap \neg C_n)^I = T^I$ iff $C_i^I = \emptyset$ for each i $(1 \leq i \leq n)$. Together with Proposition 3.2 this implies the following theorem:

Theorem 3.3:
> *A set $\{\Gamma_1,\ldots,\Gamma_n\}$ of concept constraints with pairwise disjoint variables is not valid iff there is a sequence $(x_1{:}C_1,\ldots,x_n{:}C_n)$ with $x_i{:}C_i \in \Gamma_i$ $(1 \leq i \leq n)$ such that the concept $\neg C_1 \sqcap \ldots \sqcap \neg C_n$ is universally satisfiable.*

The above result shows that we can decide validity of a finite set $\{\Gamma_1,\ldots,\Gamma_n\}$ of concept descriptions as follows: We have to test if the concept description $\neg C_1 \sqcap \ldots \sqcap \neg C_n$ is universally satisfiable for all possible choices $(x_1{:}C_1,\ldots,x_n{:}C_n)$ collecting single atomic constraints $x_i{:}C_i$ from each of the concept constraints Γ_i. The set of concept descriptions is valid, iff none of these tests returns the answer *Yes*. In the following we will show how we can decide this universal satisfiability (section 4.2). A slight modification of this algorithm provides an algorithm for the satisfiability test for concept descriptions (section 4.3).

4   Proving Universal Satisfiability and Satisfiability of Concept Descriptions

Schmidt-Schauß and Smolka (1988) give an algorithm which checks the satisfiability of concept descriptions and we shall demonstrate by an example how this algorithm works. Given a concept description C this algorithm essentially tries to construct an interpretation which interprets C as a nonempty set. If this process fails, i.e., if a contradiction occurs, the concept description is not satisfiable; otherwise the algorithm generates such an interpretation.

Let us demonstrate this by an example: In order to show that the concept description $C = \exists R.A \sqcap \forall R.B$ is valid, we want to construct a finite interpretation I such that C^I is a nonempty set, i.e., there exists an element in T^I which is an element of C^I. The algorithm generates such an element a and imposes the constraint $a \in C^I$ on it. Obviously, this yields that $a \in (\exists R.A)^I$ and $a \in (\forall R.B)^I$. From $a \in (\exists R.A)^I$ we can deduce that there has to exist an element b such that $(a, b) \in R^I$ and $b \in A^I$. Thus the algorithm introduces a new element b and constrains it as mentioned. Since $a \in (\forall R.B)^I$ and $(a, b) \in R^I$, we also get the constraint $b \in B^I$. We have now satisfied all the constraints in the example without getting a contradiction. This shows that C^I is satisfiable. We have generated an interpretation I which interprets C as a nonempty set: $\mathsf{T}^I = \{a, b\}$, $A^I = B^I = \{b\}$, and $R^I = \{(a, b)\}$. Termination of the algorithm is ensured by the fact that the newly introduced constraints are always smaller than the constraints which enforced their introduction. Note that a constraint $a \in (C \sqcap D)^I$ forces to generate two new constraints $a \in C^I$ and $a \in D^I$. On the other hand, if we have a constraint $a \in (C \sqcup D)^I$, then we have to choose either $a \in C^I$ or $a \in D^I$.

Now let us consider an algorithm which checks the universal satisfiability of concept descriptions. Suppose $C = \exists R.A \sqcap \forall R.B$ is universally satisfiable. We try to construct an interpretation I such that $C^I = \mathsf{T}^I$. Since T^I is a nonempty set, there has to exist an element in T^I, and hence in C^I. Thus the algorithm generates such an element a and imposes the constraint $a \in C^I$ on it. Furthermore, as argued above, the algorithm creates the constraints $(a, b) \in R^I$, $b \in A^I$, $b \in B^I$ where b is a newly introduced element. Since b is an element of T^I, we have to impose the additional constraint $b \in C^I$ on b. Again, because $b \in C^I$, there exists an element $c \in \mathsf{T}^I$ which satisfies the constraints $(b, c) \in R^I$, $c \in A^I$, $c \in B^I$, $c \in C^I$. Obviously, the algorithm that tries to construct an interpretation I with $C^I = \mathsf{T}^I$ would run forever creating more and more new elements. However, if we set $c = b$, then we have $\mathsf{T}^I = \{a, b\}$, $A^I = B^I = \{b\}$, $R^I = \{(a, b), (b, b)\}$, and hence $C^I = \mathsf{T}^I$. Thus, the main problem in extending the satisfiability algorithm to a universal satisfiability algorithm is to find an appropriate criterion for termination. This will be done with the help of so-called concept trees, which are introduced in the next subsection.

This section is organized as follows: In subsection 4.1 we state some basic definitions. The main result, an algorithm for deciding the universal satisfiability of concept descriptions, is presented in subsection 4.2. In subsection 4.3 we slightly modify this algorithm to obtain an algorithm for checking the satisfiability of concept descriptions.

4.1 Basic Definitions

To keep our algorithms simple, we single out a special class of concept descriptions as normal forms. We say a concept description C is **equivalent** to D iff $C^I = D^I$ for every interpretation I. A concept description C is called **simple** iff C is a concept symbol, or a complemented concept symbol, or if C is of the form $\forall R.D$ or $\exists R.D$. A **conjunctive** concept description has the form $C_1 \sqcap ... \sqcap C_n$ where each C_i is a simple concept description. A **subconjunction** for $C_1 \sqcap ... \sqcap C_n$ has the form $C_{i_1} \sqcap ... \sqcap C_{i_m}$. By grouping together exists and value restrictions we can write conjunctive concept descriptions in the form
$$A_1 \sqcap ... \sqcap A_m \sqcap \exists R_1.E_1 \sqcap ... \sqcap \exists R_l.E_l \sqcap \forall S_1.D_1 \sqcap ... \sqcap \forall S_k.D_k.$$
This concept description contains a **clash** iff there exist A_i and A_j such that $A_i = \neg A_j$, and contains an **exists-restriction** iff $l > 0$. A **disjunctive** concept description has the form $C_1 \sqcup ... \sqcup C_n$ where each C_i is a conjunctive concept description. A disjunctive concept

description which is equivalent to a concept description C is called a ***disjunctive normal form*** for C.

Every concept description can be transformed into a disjunctive normal form. This transformation can be performed as follows: First, we compute the ***negation normal form*** of a concept description, that is, we bring the negation signs immediately in front of concept symbols by rewriting the concept description via de Morgan's laws and with rules $\neg \forall R.C \Rightarrow \exists R.\neg C$ and $\neg \exists R.C \Rightarrow \forall R.\neg C$. Then we transform this concept description into a disjunctive normal form by applying the associativity, commutativity, idempotency and distributivity laws of conjunction and disjunction.

We now define concept trees, which are used to impose a control structure on the algorithm. A ***directed graph*** $G = (N, E)$ consists of a (not necessarily finite) set of nodes N and a set of edges $E \subseteq N \times N$. A ***path*** in a directed graph is a sequence $N_1,\dots, N_k$ of nodes such that (N_i, N_{i+1}) is an edge for each i, $1 \leq i < k$. Notice that paths contain at least two different nodes. We say that this path is a path ***from*** N_1 ***to*** N_k, N_1 is a ***predecessor*** of N_k and N_k is a ***successor*** of N_1. For a path consisting of two nodes N, M we say N is a ***direct predecessor*** of M and M is a ***direct successor*** of N. A node without successors is a ***leaf***. A ***tree*** is a directed graph such that

➤ there is one node, called the ***root***, that has no predecessor

➤ each node other than the root has exactly one predecessor.

A ***concept tree*** is a tree such that every node is equipped with components, namely

➤ *type*

➤ *extended*

➤ *concept-description*

➤ *value.*

The values for the component *type* range over the symbols "$\sqcap$", "$\sqcup$" and "$\exists R$" where R is a role symbol, for the component *extended* they range over the symbols "yes" and "no", and for *value* they range over the symbols "solved", "clash", "cycle" and "null". The values for the component *concept-description* are concept descriptions. Given a node N in a concept tree we will access the contents of the corresponding component with N.component. Figure 4.1 shows a concept tree.

A concept tree T is called ***extended*** if for every node N in T N.extended = "yes".

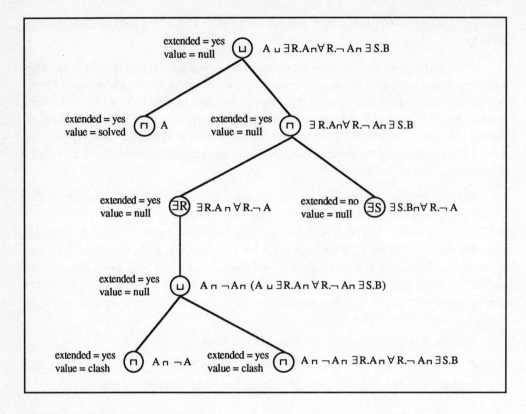

Figure 4.1: A concept tree.

4.2 An Algorithm for Deciding the Universal Satisfiability of Concept Descriptions

We will now present an algorithm that decides whether a given concept description is universally satisfiable. The algorithm proceeds as follows: First, a concept tree consisting of a single node is created. Then, in successive propagation steps new nodes are added until we obtain an extended concept tree. We will see that the given concept description is universally satisfiable if and only if the extended concept tree satisfies a certain condition, which can be checked easily.

The algorithm uses several functions. The function Universal-Satisfiability takes a concept description as input, creates a concept tree, and returns this tree as argument to the function Extend-concept-tree. This function expands the concept tree by iterated calls of the functions Expand-or-node, Expand-and-node, and Expand-∃R-node until an expanded concept tree is obtained.

The function **Universal-Satisfiability** takes a concept description C as input and creates a concept tree T. This concept tree consists of the node root with

➤ root.type = "⊔"

➤ root.extended = "no"

➤ root.value = "null"   and

root.concept-description contains a disjunctive normal form for C. Then the function Extend-concept-tree is called with T as argument.

The function **Extend-concept-tree** takes a concept tree as argument and returns an extended concept tree. It uses the functions Extend-or-node, Extend-and-node, and Extend-∃R-node as subfunctions. Here is the formulation of the function Extend-concept-tree in a Pascal-like notation.

Algorithm Expand-concept-tree (T)

    if T is extended

        then return T

    elsif T contains a node N such that N.type = "⊔" and N.extended = "no"

        then Expand-or-node(T,N)

    elsif T contains a node N such that N.type = "⊓" and N.extended = "no"

        then Expand-and-node(T,N)

    else let N be a node in T such that N.type = "∃R" and N.extended = "no"

        Expand-∃R-node(T,N)

end Expand-concept-tree.

The function **Expand-or-node** takes a concept tree T and a node N occurring in T as arguments and returns a concept tree T'. Suppose $C_1 \sqcup C_2 \sqcup \dots \sqcup C_n$ is a disjunctive normal form of N.concept-description. We modify T (and obtain T') such that N.extended = "yes" and the (newly created) nodes N_i, $1 \leq i \leq n$, with

➤ N_i.type = "⊓"

➤ N_i.extended = "no"

➤ N_i.concept-description = C_i

➤ N_i.value = "null"

are successors of N.

The function **Expand-and-node** takes a concept tree T and a node N occurring in T as arguments and returns a concept tree T'. We modify T (and obtain T') such that N.extended = "yes" and N.value is

➤ "solved"   if N.concept-description doesn't contain an exists-restriction

➤ "clash"   if N.concept-description contains a clash

➤ "cycle" if there is a predecessor N' of N such that N'.type = "$\sqcap$" and N'.concept-description is equal to N.concept-description modulo associativity, commutativity, and idempotency

➤ "null" otherwise.

Furthermore, if N.value = "null", we create successors for N in the following way. Suppose N.concept-description = $A_1 \sqcap ... \sqcap A_n \sqcap \exists R_1.C_1 \sqcap ... \sqcap \exists R_l.C_l \sqcap \forall S_1.D_1 \sqcap ... \sqcap \forall S_k.D_k$. Then for every i, $1 \le i \le l$, the (newly created) node N_i with

➤ N_i.type = "$\exists R_i$"

➤ N_i.extended = "no"

➤ N_i.concept-description = $A_1 \sqcap ... \sqcap A_n \sqcap \exists R_i.C_i \sqcap \forall S_1.D_1 \sqcap ... \sqcap \forall S_k.D_k$

➤ N_i.value = "null"

is a successor of N.

The function **Expand-$\exists$R-node** takes a concept tree T and a node N occurring in T as arguments and returns a concept tree T'. Suppose N.concept-description = $A_1 \sqcap ... \sqcap A_n \sqcap \exists R.C \sqcap \forall S_1.D_1 \sqcap ... \sqcap \forall S_k.D_k$. We modify T (and obtain T') such that N.extended = "yes" and the (newly created) node N' with

➤ N'.type = "$\sqcup$"

➤ N'.extended = "no"

➤ N'.concept-description = root.concept-description $\sqcap C \sqcap \prod_{R = S_j, 1 \le j \le k} D_j$

➤ N'.value = "null"

is a successor of N.

Let C be a finite concept description. Obviously, C contains finitely many subterms. Hence $\{ \prod_{1 \le i \le n} C_i \mid C_i$ is a subterm of $C\}$ contains finitely many elements modulo associativity, commutativity, and idempotency.

Proposition 4.1: (Termination)

Let C be a concept description. Then Universal-Satisfiability(C) terminates.

Proof: Assume that the algorithm Universal-Satisfiability does not terminate. Then an infinite concept tree is generated since each call of Extend-or-node, Extend-and-node, or Extend-$\exists$R-node adds new nodes to the concept tree. Since every node has finitely many successors we conclude with König's Lemma that there exists an infinite path in this tree. This infinite path contains infinitely many nodes $N_1, N_2, ...$ with N_i.type = "$\sqcap$" and N_i.concept-description is not equal to N_j.concept-description modulo idempotency and commutativity for $i < j$; or otherwise we would have then N_j.value = "cycle" and thus N_j would be a leaf. On the other hand, it is easy to show that any concept description of a node N with N.type = "$\sqcap$" is of the form $C_1 \sqcap ... \sqcap C_n$ where the C_i are subterms of root.concept-description. Since root.concept-description is finite there exist only finitely many concept descriptions modulo associativity, commutativity, and idempotency that are conjunctions of subterms of root.concept-description as mentioned before. Thus, there cannot exist an infinite path which implies that the algorithm terminates. ❑

An *instance* is obtained from a concept tree by keeping for a node N with N.type = "$\sqcap$" or N.type = "$\exists R$" all direct successors, and by keeping for a node N with N.type = "$\sqcup$" only one of its direct successors. Figure 4.2 shows the instances for the concept tree given in Figure 4.1.

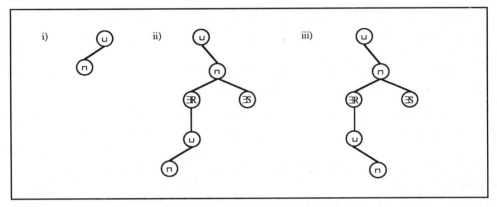

Figure 4.2: The three instances of the concept tree given in Figure 4.1.

An instance T is **successful** iff for every leaf N in T N.value = "solved" or N.value = "cycle".

Proposition 4.2: (Completeness)
Let C be a concept description and let T be the extended concept tree computed by Universal-Satisfiability(C). If C is universally satisfiable, then there exists a successful instance of T.

Proof: Let C be a concept description and let I be an interpretation such that $C^I = T^I$. Furthermore, let T be the extended concept tree computed by Universal-Satisfiability(C). Then root.concept-description = C. We use I to construct a successful instance of T. That means, for every node N with N.type = "$\sqcup$" we have to choose exactly one direct successor.

Suppose $a \in T^I$. The root is a node of type "$\sqcup$" and we have an element $a \in C^I$. If $C_1 \sqcup \dots \sqcup C_n$ is a disjunctive normal form for C, then there exist direct successors $N_1, \dots, N_n$ of root such that N_i.type = "$\sqcap$" and N_i.concept-description = C_i for i, $1 \leq i \leq n$. Since $a \in (C_1 \sqcup \dots \sqcup C_n)^I$ there exists an i such that $a \in C_i^I$. We choose the node N_i as direct successor for the root in our instance. Since $a \in C_i^I$, there is no clash in C_i and hence N_i.value $\neq$ "clash". If N_i.value = "solved" or N_i.value = "cycle", then N_i doesn't have any successor and we are done. Otherwise, if N_i.value = "null", consider the concept description C_i, which has the form $A_1 \sqcap \dots \sqcap A_m \sqcap \exists R_1.E_1 \sqcap \dots \sqcap \exists R_l.E_l \sqcap \forall S_1.D_1 \sqcap \dots \sqcap \forall S_k.D_k$. Since $a \in C_i^I$ we have $a \in \exists R_i.E_i^I$ for i, $1 \leq i \leq l$. There exists $b_i \in T^I$ such that $(a, b_i) \in R_i^I$ and $b_i \in E_i^I$ ($1 \leq i \leq l$.)

Furthermore $b_i \in D_j$, if $R_i = S_j$ $(1 \leq j \leq k)$ since $a \in (\forall S_j.D_j)^I$. Since $b_i \in \mathsf{T}^I$, we also have $b_i \in C^I$. Hence $b_i \in (C \sqcap E_i \sqcap \bigsqcap_{R_i = S_j} D_j)^I$. By construction of the concept tree the node N_i has exactly l direct successors $M_1, ..., M_l$. Every M_i has exactly one direct successor M_i' with

➤ M_i'.concept-description $= C \sqcap E_i \sqcap \bigsqcap_{R_i = S_j} D_j$

➤ M_i'.type $= $ "$\sqcup$".

and for each i we have an element $b_i \in (C \sqcap E_i \sqcap \bigsqcap_{R_i = S_j} D_j)^I$. We can now proceed with these nodes of type "$\sqcup$" as described above for root. Since the extended concept tree T computed by Universal-Satisfiability(C) is finite, this construction process terminates. Note that the constructed instance does not contain a node N with N.value $=$ "clash" and hence we have a successful instance.

Thus we have shown that, given a universally satisfiable concept description C, the expanded concept tree contains a successful instance. ❑

Let S be a successful instance of the extended concept tree computed by Universal-Satisfiability(C). Then S yields a *canonical interpretation* I, which is defined as follows.

(1) The elements of the domain T^I are the nodes $N_1, ..., N_n$ in S such that N_i.type $=$ "$\sqcap$" and N_i.value $\neq$ "cycle".

(2) Interpretation of role symbols: Let N be an element of T^I. Then N.type $=$ "$\sqcap$" and N.value $=$ "solved" or N.value $=$ "null". If N.value $=$ "solved", then N is a leaf in S, and for any role R, N does not have an R-successor in I. If N.value $=$ "null", then there exist direct successors $M_1, ..., M_n$ of N with M_i.type $=$ "$\exists R_i$" for i, $1 \leq i \leq n$. Every M_i has exactly one direct successor M_i' with M_i'.type $=$ "$\sqcup$", and every M_i' has exactly one direct successor M_i'' with M_i''.type $=$ "$\sqcap$". If M_i''.value $=$ "solved" or M_i''.value $=$ "null", then we set $(N, M_i'') \in R_i^I$. Note that $M_i'' \in \mathsf{T}^I$. Otherwise M_i''.value $=$ "cycle" and there exists a predecessor P of M_i'' such that P.type $=$ "$\sqcap$", P.value $\neq$ "cycle", and P.concept-description is equal to M_i''.concept-description modulo associativity, commutativity, and idempotency. In this case we set $(N, P) \in R_i^I$. Note that $P \in \mathsf{T}^I$.

(3) Interpretation of concept symbols: Suppose $N \in \mathsf{T}^I$. Then N.concept-description is of the form $A_1 \sqcap ... \sqcap A_m \sqcap \exists R_1.E_1 \sqcap ... \sqcap \exists R_l.E_l \sqcap \forall S_1.D_1 \sqcap ... \sqcap \forall S_k.D_k$. If A_i $(1 \leq i \leq m)$ is a non-negated concept symbol, then we set $N \in A_i^I$. Note that N.value $\neq$ "clash", and hence there does not exist an A_j with $A_j = \neg A_i$. Thus $N \in (A_1 \sqcap ... \sqcap A_m)^I$.

Before we can show that this canonical interpretation satisfies $\mathsf{T}^I = C^I$, we need one more definition and an observation concerning concept descriptions of nodes with type "$\sqcap$".

The **depth** τ of a concept description in negation normal form is defined as follows.

➤ $\tau(C) = \tau(\neg C) = 0$ if C is a concept symbol

➤ $\tau(\forall R.C) = \tau(\exists R.C) = 1 + \tau(C)$

➤ $\tau(C \sqcap D) = \tau(C \sqcup D) = max \{\tau(C), \tau(D)\}$.

Let C be a concept description and let T be the extended concept tree computed by Universal-Satisfiability(C). If N is a node in T with N.type = "$\sqcup$", then N.concept-description has the form $C \sqcap C_{rest}$. Let $C_1 \sqcup ... \sqcup C_n$ be a disjunctive normal form of $C \sqcap C_{rest}$. There exist direct successors $N_1, ..., N_n$ of N with N_i.type = "$\sqcap$" and N_i.concept-description = C_i. Obviously, C subsumes $C \sqcap C_{rest}$, and $C \sqcap C_{rest}$ subsumes each C_i. Thus, if N_i is a node in T with N_i.type = "$\sqcap$", then root.concept-description (which is C) subsumes N_i.concept-description.

Proposition 4.3: (Soundness)

Let C be a concept description and let T be the extended concept tree computed by Universal-Satisfiability(C). If there exists a successful instance of T, then C is universally satisfiable.

Proof: Let C be a concept description, let T be the extended concept tree computed by Universal-Satisfiability(C), and let S be a successful instance of T. Note that root.concept-description = C. We will show that the canonical interpretation I induced by S interprets C as T^I. First we prove the following claim.

Claim: Let I be the canonical interpretation induced by S and let $N \in T^I$. If D is a subconjunction of N.concept-description, then $N \in D^I$.

Proof: (by **induction on the depth** $\tau(D)$.) Let $N \in T^I$. Then N is a node in S with N.type = "$\sqcap$" and N.value $\neq$ "cycle". We know that N.concept-description is of the form $A_1 \sqcap ... \sqcap A_m \sqcap \exists R_1.E_1 \sqcap ... \sqcap \exists R_l.E_l \sqcap \forall S_1.D_1 \sqcap ... \sqcap \forall S_k.D_k$. Now let D be a subconjunction of N.concept-description.

$\tau(D) = 0$. Then D is of the form $A_{i_1} \sqcap ... \sqcap A_{i_n}$. By definition of the interpretation of concept symbols we have $N \in (A_1 \sqcap ... \sqcap A_m)^I$, which implies $N \in (A_{i_1} \sqcap ... \sqcap A_{i_n})^I$.

$\tau(D) > 0$. Let $D = D_1 \sqcap ... \sqcap D_n$. We have to show for all i, $1 \leq i \leq n$, that $N \in D_i{}^I$. If $\tau(D_i) < \tau(D)$ we know by the induction hypothesis $N \in D_i{}^I$. Now suppose $\tau(D_i) = \tau(D)$. Then D_i is of the form $\forall R.E$ or $\exists R.E$, where $\tau(E) = \tau(D) - 1$.

(1) Suppose $D_i = \forall R.E$. We have to show that for any $M \in T^I$, $(N, M) \in R^I$ implies $M \in E^I$. Consider the node N. If $(N, M) \in R^I$, then there exists a direct successor N' of N with N'.type = "$\exists R$". The node N' has a direct successor N'' with N''.type = "$\sqcup$". Furthermore, N''.concept-description = $C \sqcap E \sqcap \lceil\rceil_{R = S_j} D_j$. Suppose $E_1 \sqcup ... \sqcup E_n$ is a disjunctive normal form of E. It is easy to see that, if $C_1 \sqcup ... \sqcup C_m$ is a disjunctive normal form of N''.concept-description, then every C_i contains an E_j as subconjunction for some j, $1 \leq j \leq n$. Since N''.type = "$\sqcup$", N'' has exactly one direct successor P with P.type = "$\sqcap$" and P.concept-description = C_i for some i, $1 \leq i \leq m$. If P.value $\neq$ "cycle", P is equal to M. Otherwise P.value = "cycle", and M is a predecessor of N'' with M.type = "$\sqcap$" and M.concept-description is equal to C_i modulo idempotency and commutativity. As mentioned before E_j is a subconjunction of C_i. Since $\tau(E_j) \leq \tau(E) < \tau(\forall R.E)$ we conclude by induction that $M \in E_j{}^I$ and hence $M \in E^I$. Thus we have shown that $N \in (\forall R.E)^I$.

(2) Suppose $D_i = \exists R.E$. We have to show that there exists an $M \in T^I$ with $(N, M) \in R^I$ and $M \in E^I$. Consider the node N. Since $\exists R.E$ is a subconjunction of N.concept-description there exists a direct successor N' of N with N'.type $=$ "$\exists R$". Furthermore, N' has a direct successor N'' with N''.type $=$ "$\sqcup$" and N''.concept-description $= C \sqcap E \sqcap \lceil_R = s_j D_j$. Suppose $E_1 \sqcup \ldots \sqcup E_n$ is a disjunctive normal form of E. As shown in (1) we know that, for every direct successor M of N'', there exists j, $1 \leq j \leq n$, such that E_j is a subconjunction of M.concept-description. Note that M.type $=$ "$\sqcap$". If M.value $\neq$ "cycle", then $M \in T^I$ and by definition of I we have $(N, M) \in R^I$. Since $\tau(E_j) < \tau(\forall R.E)$ we conclude by induction that $M \in E_j^I$ and hence $M \in E^I$. Otherwise, if M.value $=$ "cycle", then there exists a predecessor M' of M with M'.type $=$ "$\sqcap$" and M'.concept-description is equal to M.concept-description modulo idempotency and commutativity. By definition of I we know that $M' \in T^I$ and $(N, M') \in R^I$. As before we conclude by induction that $M' \in E_j^I$ and hence $M' \in E^I$. Thus we have shown that $N \in (\exists R.E)^I$. ❏

Let I be the canonical interpretation induced by S and let $N \in T^I$. We have shown that, if D is a subconjunction of N.concept-description, then $N \in D^I$. Every concept description is a subconjunction of itself and hence $N \in (N.\text{concept-description})^I$. Since N.type $=$ "$\sqcap$", we know that C subsumes N.concept-description (as mentioned above). Hence for every $N \in T^I$ we have $N \in C^I$. Thus $T^I = C^I$. ❏

Let C be a concept description. We have shown that the call Universal-Satisfiability(C) terminates and returns an expanded concept tree T (Proposition 4.1). If C is universal satisfiable, then T contains a successful instance (Proposition 4.2), and if T contains a successful instance, then C is universally satisfiable (Proposition 4.3). Thus we have proven the main result of this paper: (?)

Theorem 4.4:

  A concept description C is universally satisfiable if and only if the extended concept tree computed by Universal-Satisfiability(C) contains a successful instance.

Note that it can be easily decided whether a concept tree does contain a successful instance by using a depth-first search. For further remarks about the implementation of the algorithm see subsection 4.4.

4.3 Deciding the Satisfiability of Concept Descriptions

As mentioned before (Schmidt-Schauß & Smolka, 1988) have already described an algorithm for deciding the satisfiability of concept descriptions. However, such an algorithm can also be obtained within our formalism developed for deciding universal satisfiability. Since this requires only a slight modification of the function Universal-Satisfiability, we will include a description of this algorithm. The idea underlying the modification is as follows. Suppose we want to decide whether a concept description C_0 is universally satisfiable. The function Universal-Satisfiability tries to construct an interpretation I such that $T^I = C_0^I$ and $T^I \neq \emptyset$. To do this the function generates an element which is in T^I. In general, constraints imposed on this element force the function to generate further elements which are in T^I. Since we want to construct an interpretation with $T^I = C_0^I$, the newly introduced elements also have to be in C_0^I.

A satisfiability algorithm has to do less work. Since we only want to find an interpretation I such that C_0^I is nonempty, it is sufficient to guarantee that there exists one element which is in C_0^I. Thus, there is no need to force the newly introduced elements to be in C_0^I.

This observation leads us to the following straightforward modification of the function Extend-∃R-node used in Universal-Satisfiability. Let N be a node with N.type = "∃R" and N.concept-description = $A_1 \sqcap \ldots \sqcap A_n \sqcap \exists R.C \sqcap \forall S_1.D_1 \sqcap \ldots \sqcap \forall S_k.D_k$. The function Extend-∃R-node creates a new node N' such that N'.concept-description = root.concept-description $\sqcap C \sqcap \bigsqcap_{R = S_j, 1 \leq j \leq k} D_j$. Recall that the initialization step of the algorithm was done in a way such that root.concept-description = C_0. Thus we just have to omit root.concept-description in the definition of N'.concept-description, i.e., we modify Extend-∃R-node such that N'.concept-description = $C \sqcap \bigsqcap_{R = S_j, 1 \leq j \leq k} D_j$. We obtain the function **Satisfiability** from Universal-Satisfiability by taking this modified version of Extend-∃R-node.

Let C_0 be a concept description and let T be an extended concept tree computed by Satisfiability(C_0). Furthermore, let N, M be nodes in T with N.type = M.type = "$\sqcap$". It is easy to see that $\tau(N$.concept-description$) > \tau(M$.concept-description$)$ if N is a predecessor of M. As a consequence, N.value ≠ "cycle" for every node N in T. Furthermore, if $\tau(C_0) = n$, the longest path in S contains at most n nodes of type "$\sqcap$", and hence at most $3 * n$ nodes. Since obviously every node in S has finitely many direct successors we have the following result.

Proposition 4.5: (Termination)
Let C_0 be a concept description. Then Satisfiability(C_0) terminates without creating a node with value "cycle".

In the following we will prove that a concept description C_0 is satisfiable if and only if there exists a successful instance in the extended concept tree computed by Satisfiability(C_0). As in Chapter 4.2 one can show the completeness (Proposition 4.6) and the soundness (Proposition 4.7) of the algorithm.

Proposition 4.6: (Completeness)
Let C_0 be a concept description and let T be the extended concept tree computed by Satisfiability(C_0). If C_0 is satisfiable, then there exists a successful instance of T.

Proof: Easy modification of the proof of Proposition 4.2. ❑

Proposition 4.7: (Soundness)
Let C_0 be a concept description and let T be the extended concept tree computed by Satisfiability(C_0). If there exists a successful instance of T, then C_0 is satisfiable.

Proof: Let C_0 be a concept description and let S be a successful instance of the extended concept tree computed by Satisfiability(C_0). Consider the canonical interpretation I induced by S and $N \in T^I$. We will show that C_0^I is a nonempty set. To that purpose we will use the following claim:

Claim: Let I be the canonical interpretation induced by S and let $N \in T^I$. Then $N \in$ (N.concept-description)^I. As mentioned above, if N and M are elements of T^I, then τ(N.concept-description) $> \tau$(M.concept-description) if N is a predecessor of M. Thus one can easily prove the claim by **induction on the depth** τ(N.concept-description). The proof of this claim is similar to the proof of the claim in Proposition 4.3. However it is easier for the following reason. In the proof of Proposition 4.3 we had to formulate a stronger induction hypothesis, since if N is predecessor of M, then τ(N.concept-description) is not necessarily greater than τ(M.concept-description).

Let $C_1 \sqcup ... \sqcup C_n$ be a disjunctive normal form of C_0. Then there exists a node in S such that N.type = "$\sqcap$" and N.value $\neq$ "cycle". Hence $N \in T^I$. We conclude that $N \in C_i^I$ and $N \in C_0^I$. Thus we have shown that C_0 is satisfiable. ❑

5 Conclusion

In the previous two sections we described algorithms for deciding satisfiability and universal satisfiability. These algorithms were presented for theoretical purposes, i.e. for proving soundness and completeness, rather than for an actual implementation. The algorithms as presented above suffer from two main sources of unnecessary complexity: The first point is that we construct a complete concept tree before testing for the existence of a successful instance. As soon as a successful instance is found the remaining unsearched part of the tree was constructed in vain. Thus an actual implementation should combine the generation of the tree with searching for a successful instance. This can be realized by a simple depth first strategy with backtracking to a previous "or"-node if a clash is encountered. Consequently, only one path has to be stored instead of keeping the whole tree in memory. As a second point, the size of a disjunctive normal form $C_1 \sqcup ... \sqcup C_n$ for a concept description C may be exponential in the size of C. In addition, if C_i leads to a successful instance, the computation of $C_{i+1}, ..., C_n$ was superfluous. However, by choosing an appropriate data structure one can enumerate the disjuncts C_i one after the other using polynomial space.

A constrained resolution prover (CORE) has been implemented along these lines in the WINO-project of the German national institute for artificial intelligence (DFKI) by our research team. CORE will be the heart of a first prototype of a knowledge representation and reasoning system (KRIS) to be developed in the WINO-project. Thereby, our future research is concentrated on following problems:

➤ As mentioned in section 2, the input formulae have to be in clause form. Since this is a strong restriction, we examine how to transform arbitrary constrained formulae into constrained clauses in the case of our application with concept constraints.

➤ In this paper we considered the concept language $\mathcal{ALC}$ as constraint theory. However, other language constructs such as number restrictions, intersection of roles etc., are used to describe taxonomical knowledge. Hence we will use such an enriched concept language as constraint theory. There are already some results how to devise algorithms for checking satisfiability and universal satisfiability in these languages.

➤ Besides concept languages, so-called assertional languages are employed in KL-ONE systems to represent knowledge about individuals. Thus, we are going to amalgamate our constraint theory by allowing such assertions.

Acknowledgements. We would like to thank Erich Achilles, Armin Laux, Jörg Peter Mohren, and Thomas Steinbach for their implementational work.
This research was supported by the German Bundesministerium für Forschung und Technologie under grant ITW 8903 0 and by the Esprit Basic Research Project 3012, COMPULOG.

References

Baader, F.: *"Regular Extensions of KL-ONE"*, DFKI Research Report, forthcoming.

Beierle, C., Hedtstück, U.,Pletat, U., Schmitt, P. H., Siekmann, J.: *"An Order-Sorted Logic for Knowledge Representation Systems"*, IWBS Report 113, IBM Deutschland, 1990.

Brachman, R. J., Schmolze, J. G.: "An Overview of the KL-ONE Knowledge Representation System", *Cognitive Science*, 9(2), pp. 171-216, 1985.

Brachman, R.J., Levesque, H.J.: *Readings in Knowledge Representation*. Morgan Kaufmann Publishers, 1985.

Brachman, R.J., Pigman Gilbert, V., Levesque, H.J.: "An Essential Hybrid Reasoning System: Knowledge and Symbol Level Account in KRYPTON", in *Proceedings of the 9th IJCAI*, pp. 532-539, Los Angeles, Cal., 1985.

Bürckert, H.-J.: *"A Resolution Principle for a Logic with Restricted Quantifiers"*, Dissertation, Universität Kaiserslautern, Postfach 3049, D-6750 Kaiserslautern, West-Germany, 1990.

Bürckert, H.-J.: "A Resolution Principle for Clauses with Constraints", in *Proceedings of 10th International Conference on Automated Deduction*, Springer LNAI 449, pp. 178-192, 1990.

Cohn, A. G.: " A More Expressive Formulation of Many-Sorted Logic", *JAR* 3,2, pp. 113-200, 1987.

Frisch, A.: "A General Framework for Sorted Deduction: Fundamental Results on Hybrid Reasoning", in *Proceedings of International Conference on Principles of Knowledge Representation and Reasoning*, pp. 126-136, 1989.

Frisch, A.: "An Investigation into Inference with Restricted Quantification and Taxonomic Representation", *Logic Programming Newsletters*, 6, pp. 5-8, 1986.

Höhfeld, M., Smolka, G.: *"Definite Relations over Constraint Languages"*, LILOG-Report 53, IBM Deutschland, 1988.

Hollunder, B.: "Hybrid Inferences in KL-ONE-based Knowledge Representation Systems", DFKI Research Report RR-90-06, DFKI, Postfach 2080, D-6750 Kaiserslautern, West-Germany, 1990. Also in the *Proceedings of the 14th German Workshop on Artificial Intelligence*, Springer-Verlag, 1990.

Hollunder, B., Nutt, W.: "Subsumption Algorithms for Concept Languages", DFKI Research Report RR-90-04, DFKI, Postfach 2080, D-6750 Kaiserslautern, West-Germany, 1990. Also in the *Proceedings of the 9th European Conference on Artificial Intelligence*, Pitman Publishing, 1990.

Huet, G.: *"Constrained Resolution - A Complete Method for Higher Order Logic"*, Ph.D. Thesis, Case Western University, 1972.

Jaffar, J., Lassez, J.-L.: "Constrained Logic Programming", *Proceedings of ACM Symp. on Principles of Programming Languages*, 111-119, 1987.

Levesque, H. J., Brachman, R. J.: "Expressiveness and Tractability in Knowledge Representation and Reasoning", *Computational Intelligence*, 3, pp. 78-93, 1987.

MacGregor, R., Bates, R.: *"The Loom Knowledge Representation Language"*, Technical Report ISI/RS-87-188, University of Southern California, Information Science Institute, Marina del Rey, Cal., 1987.

Nebel, B.: *"Reasoning and Revision in Hybrid Representation Systems"*, Lecture Notes in Artificial Intelligence, LNAI 422, Springer-Verlag, 1990.

Oberschelp, A.: "Untersuchungen zur mehrsortigen Quantorenlogik", *Mathematische Annalen*, 145:297-333,1962.

Quillian, R. M.: "Semantic Memory", in *Semantic Information Processing* (ed. M. Minsky), pp. 216-270, MIT-Press, 1968.

Robinson, J.A.: "A Machine Oriented Logic Based on the Resolution Principle", *J. ACM*, 12(1), pp. 23-41, 1965.

Schmidt, A.: "Ueber deduktive Theorien mit mehreren Sorten von Grunddingen", *Mathematische Annalen*, 115:485-506,1938.

Schmidt, A.: "Die Zulässigkeit der Behandlung mehrsortiger Theorien mittels der üblichen einsortigen Prädikatenlogik", *Mathematische Annalen*, 123:187-200,1951.

Schmidt-Schauß, M., Smolka, G.: *"Attributive Concept Descriptions with Unions and Complements"*, SEKI Report SR-88-21, FB Informatik, University of Kaiserslautern, West Germany, 1988. To appear in Artificial Intelligence.

Schmidt-Schauß, M.: *"Computational Aspects of an Order-sorted Logic with Term Declarations"*, Lecture Notes on Artificial Intelligence, LNAI 395, Springer, 1989

Smolka, G.: *"Logic Programming over Polymorphically Order-sorted Types"*, Dissertation, Universität Kaiserslautern, 1989.

Stickel, M. E.: "Automated Deduction by Theory Resolution", *Journal of Automated Reasoning*, 1:333-355, 1985.

Vilain., M. B.: "The Restricted Language Architecture of a Hybrid Representation System", in *Proceeding of the 9th IJCAI*, pp. 547-551, Los Angeles, Cal., 1985.

Walther, C.: "A Many-sorted Calculus Based on Resolution and Paramodulation", *Research Notes in Artificial Intelligence,* Pitman, Morgan Kaufman Publishers, 1987.

Walther, C.: "Many-Sorted Unification", *J. ACM*, 35(1), pp. 1-17, 1988.

Weidenbach, C., Ohlbach H.-J.: "A Resolution Calculus with Dynamic Sort Structures and Partial Functions", *Proceedings of the 9th European Conference on Artificial Intelligence*, Pitman Publishing, 1990.

Programming in 2010?
A scientific and industrial challenge

Gérard Comyn

European Computer-Industry Research Centre (ECRC)
Arabellastr. 17, D-8000 München

The development of software technology in the industrial world is
rather discouraging for researchers today. Languages and systems of-
fered by software houses and computer vendors are still very poor if
compared to the state-of-the-art in advanced research. Fortran and
Cobol are still among the most widely used programming languages,
relational databases are just starting to penetrate the market. The
breathtaking speed of transfer of new hardware technology from research labs into every day life is not at all matched by a comparable development in the software area. It is very hard to explain
this enormous discrepancy of acceptance between the hardware and the
software domain. Is it a lack of flexibility on the user side, or a
lack of insight in the importance of software by the manufacturers,
or is it just a matter of complexity? Is a new piece of software
really that much more complex than a new hardware component?

Whereas the commercial world seems to be rather indifferent about
new software concepts, education tries to keep up with research in
a much more encouraging way. Students in computer science today are
as familiar with esoteric concepts like resolution and polymorphism
as with classical notations like recursion and iteration. They learn
programming in Prolog or Smalltalk as naturally as their "ancestors"
have learnt their Fortran or Algol. How will the world of profes-
sional software production change once this new generation of programmers and designers has been integrated? Or will they be changed
by their environment and resignate sooner or later?

I don't want to continue mourning about the discouraging state of
today's commercial software scene. I would rather like to be con-
fident in the ability of users to assimilate new software much
quicker than believed by the vendors. Similarly I am very much con-
fident in the ability of research and development to face new and
exciting requirements much more quickly than commonly believed.
Whether my opinion will turn out to be justified or not will mainly

depend on the <u>ability of research to cope with the immense degree of</u>
<u>additional complexity introduced by a wide range of new heterogeneous</u>
<u>concepts and paradigms.</u> Mastering such complexity on the software
level will only be possible by means of a <u>very rigorous and disci-</u>
<u>plined intellectual penetration of the problem domains relying on</u>
<u>powerful general tools for knowledge representation and manipulation.</u>
I regard the following principles as decisive in this respect:

1. Founding any kind of programming activity on a sound and reliable
   formal basis.

2. Exploiting as much as possible the benefits of a knowledge-based,
   declarative style of representing information.

3. Investing considerable effort in much more sophisticated methods
   of inference and retrieval than known today: despite of all the
   energy already spent we are still at a very rudimentary level of
   sophistication.

4. Designing the procedural environment for manipulation of and interaction with knowledge as soundly and rigorously as the declarative
   fundamentals of knowledge representation.

Many of the activities currently under way - in particular in the
field of Computational Logic - are trying to follow these principles.
Their results are mostly at the theoretical side yet, with some re-
markable prototype implementations having already emerged.

In order to have real impact on the commercial world, however, a
convincing demonstration of the <u>practical relevance of these results</u>
is urgently needed. First encouraging steps in this direction have
been made, e.g. by applying techniques of constraint logic programming
to a wide range of "real-life" problems with surprising success. Another point is as important as being able to demonstrate the practical
value of the results: being able to communicate what has been achieved
in an intelligible style. Addressing professionals outside the research community in a language which they understand will be vital for
success or failure of our efforts.

PERSPECTIVE ON COMPUTATIONAL LOGIC

Hervé Gallaire

Bull SA France
78340 Les Clayes sous Bois

The art of programming is going to be altered in radical ways in the next 20 years. Programming, as we know it today, will become the activity of more users and of fewer "programmers" by 2010. Computational Logic will have to fit in a wider context. For applications programming, and to some extent system programming, the context is likely to be the following: applications will mainly become cooperative and distributed, and distribution will be transparent to the programmer; parallelism will be hidden; programming will rely on libraries of reusable objects, both for code and data structures, with well defined application programming interfaces, supported by powerful CASE systems; applications will be written by combining existing modules, extending and specializing them; powerful event-based languages, the so-called scripts, workflow, or rule-based languages, will provide the "programming power"; the libraries and the event languages will depend on application domains; this will generalize the spirit of 4th Generation languages; examples of such languages will be Office Procedure Languages, Resource Management Languages, Business Analysis Languages, etc.This evolution is underway, through the convergence of research in programming languages, databases, artificial intelligence, object-based technology, through the emergence of distributed systems, software engineering, and the influence of all the bodies working on standards for open systems.

Where does Computational Logic fit in this picture?

From the above, it is clear that I do not think that computational logic (CL) will have taken the world over in 2010. This mitigated position is not downplaying CL, which will have very wide influence on language research, but rather a realistic assessment of the resistance to new concepts, of the relative lack of image logic still faces today, and on the ease with which most new concepts are, when understood, recast into any other more fashionable framework.

As language, CL will be used "as such" in selected areas. Symbolic problem solving is one such domain, where the constraint based languages give a preview of the type of declarative programming that will be possible on a wider scale. Even disguised in other forms, these languages provide all necessary ingredients: unification in specific domains, non determinism of the computation; they are here to grow and stay, and application fields will vary greatly. Objects and logic will have been merged; the theory is not yet fully there, but it will be. This will provide the basis for new database systems where rules and objects will cohabit nicely. These systems will be dominant in business applications; they will offer the needed combination of expressiveness and true declarative programming (which is more than object based systems provide with methods evaluation), with the additional possibility to get extended capabilities such as automatic integrity checking and view updates in a true logic framework (where "logic" here may be one of many logic systems). CL will thus be key in a newly emerging field called "business intelligence" (EIS, intelligent spreadsheets), which combines the problem solving and the business applications. CL will be key in the event based languages, which will become declarative and require problem solving capabilities as well. They may be the best chance for the concurrent versions of CL languages to survive. Similarly CL will play an essential role in programs realising the mappings between the different distributed systems.

Progress in CL will authorise very deep analysis techniques, relying on abstract interpretation and partial evaluation techniques, making it possible to have powerful transformation and compilation techniques for pure (and impure) logic programs. Program synthesis in its generality will not be possible; rather, it will become less of a necessity thanks to the transformation techniques and to the new programming methodology. I do not believe that formal proofs will be there yet, even if type theory progresses significantly; they will be subsumed by the methodologies based around transformation techniques.

To achieve this influential role and be of practical use, CL will rely more on Logic Programming (LP) than on the other flavors of CL. And LP will become a pure language, tightly integrated with other languages. Functional programming will be embedded in LP, with evaluation techniques very similar to those which are used in constraint based languages. In the other direction, functional programming will have been extended so as to handle equational theories comparable to the constraint languages.

In order to get to this stage, we'll need work on theoretical issues and it seems to me that although the CL and LP communities are still growing, there are enough scientific interesting problems that research will continue, in many different directions. But influence in the real world, and consequently research ultimately, will happen only when enough practical work will demonstrate the benefits; I am not sure we are making enough efforts to reach these goals.

If we get there, CL (and classical programming) will have replaced most of AI technology, but will not have solved AI problems, apart from problem solving ones. Under AI influence, or others, it is very likely that CL will have evolved to include more work on non classical logic (time, hypothetical reasoning, ..). This will be a major factor of influence of CL for problems where modeling is important. In summary, the challenge still lies ahead of us. CL still holds a great promise, but let's make sure we put the right emphasis on it.

Programming in the year 2010

Robert Kowalski

Imperial College, Department of Computing,

180 Queen's Gate, London SW7 2BZ, UK

In trying to make predictions about a field so young as computing, it is useful to draw comparisons with a related and much older field such as law.

There are many parallels between computing and law, some of which have been explored in our work on the formalisation of legislation as extended logic programs [1]. The most obvious parallel is the one between programs and legislation. Similarly to the way in which programs are written to be executed by machines, legislation is written to be executed by people. Although often unclear and complex, it is ordinarily written with a precision far greater than that found in ordinary natural language, and in a style similar to logic programming.

But there are other important parallels. Like programs in computing, legislation is written to meet certain ill defined and often conflicting specifications. In law these specifications reflect the social, economic, and political objectives of the government in power at the time. As in computing, programs in the law often fail to meet their specifications, and specifications change faster than programs can be adapted to meet changing needs. For the sake of "law and order", incorrect programs are enforced upon administrators and ordinary users alike.

Such parallels suggest that, in the same way that the technology of law has made only slow progress over the millennia, the progress of computer software may be inhibited because of commitments to existing systems and because of the difficulties and unpredictability of initiating change.

Most changes that occur in the law are evolutionary rather than revolutionary. In computing too, tools that support evolutionary improvement such as object-oriented programming and formal methods, might be more successful, especially in the short term, than technologies such as logic programming that are often promoted as offering the prospect of a revolutionary advance.

I believe that logic programming and computational logic, because they are based on human logic and human thinking, are ultimately better for computing than more conventional software technologies. In the short term, however, except for niche applications, the impact of these newer logic-based technologies may depend more on their ability to coexist with and improve existing applications than on their ability to revolutionize the future.

Reference

[1]   Kowalski, R. A. and Sergot, M. J. [1990] "The use of logical models in legal program solving". Ratio Juris, Vol. 3, No. 2, pp. 201-218.

It's Past Time for Practical Computer Checked Proofs of Program Correctness

John McCarthy
Computer Science Department
Stanford University
Stanford, CA 94305

In 1961 I published my first paper on computer-checked proofs of program correctness. In it I proposed that soon buyers would not pay for programs until furnished with computer checked proofs that they met there specifications. I was careful to point out that the buyer would have to be careful that the specification reflected his real requirements. Alas, almost 30 years later computer checked proofs are still not in practical use.

Why, and what are the prospects?

In successive PhD theses, the students of Boyer and Moore have demonstrated proofs that substantial programs and substantial pieces of hardware meet their specifications. Unfortunately, that's probably not good enough. The amount of work and time that goes into a PhD thesis and the level of ability required are too much for the verification of even quite important programs.

Another problem is the specifications themselves. If they could be available at the time programming begins in a fully formal form, meeting them would be a well-defined task. Unfortunately, the buyers of programs aren't yet capable of writing formal specifications.

Still 30 years is a long time, and we ought to be further along.

I think the biggest difficulty is that the task requires abilities presently possessed only by mathematicians and logicians and also the motivation of engineers to complete a specific tedious task. Experience shows that engineers can acquire mathematical skills adequate to achieve engineering goals provided the skills are adequately taught.

There is considerable confusion about the specifications of programs that interact with the outside world rather than merely providing an answer. As discussed in my previous paper in this symposium, one needs to distinguish between two kinds of specifications, which I called *illocutionary* and *perlocutionary*, the one concerning input-output relations of the program and the other what the program accomplishes in the world. Verification of the first involves only facts of computer science, while the latter can be done only on the basis of assumptions about the world.

I think the people who claim that formal program verification cannot be a practical technique are mistaken, and one source of their mistake is confusion about the two kinds of specification and their dependence on assumptions about the world. The assumptions about the world are not themselves subject to formal verification, except sometimes on the basis of other assumptions. Nevertheless, we are always entrusting our lives to such assumptions, and with reasonable success.

Finally, the program verification community is seriously at fault in neglecting the task of developing a real technology on the basis of their mathematical achievements. There isn't sufficient comparison of the problems people work on with what is required for technology. For this reason, purely mathematical goals are too dominant.

Computational Logic needs Symbolic Mathematics

Dana S. Scott
Hillman University Professor of
Computer Science, Mathematical Logic, and Philosophy
Carnegie Mellon University
Pittsburgh, Pennsylvania

In a list entitled "Epigrams on Programming" published a few years ago, the late and well known computer pioneer Alan J. Perlis said (Epigram 65):

> Make no mistake about it: Computers process numbers – not symbols. We measure our understanding (and control) by the extent to which we can arithmetize an activity.

True as this may have been at one time, there is absolutely no reason today not to think of computers as symbol processors. The big question is how to use the powers that are available to us most effectively. Perlis also said (Epigram 36):

> The use of a program to prove the four-color theorem will not change mathematics – it merely demonstrates that a theorem, a challenge for a century, is probably not important to mathematics.

I simply cannot agree with this statement either, and I can point to considerable evidence to the contrary. I do wish, however, Professor Perlis were still with us to debate these questions, since he would not only make us sweat to defend ourselves, but he would also have new epigrams to raise our collective blood pressure and our awareness of the issues.

My recent experience with symbolic computation systems, and my investigation of some of the current proof systems, has lead me to the (pretty obvious) conclusion that there cannot be one monolithic system for computational logic. In the first place, our machines are changing too quickly, and systems will be changing all the time. In the second place, there are many projects and research groups proceeding in parallel, and they will always be coming up with new facilities. In the third place, there are many different sets of interests on different kinds of problems. It is doubtful that everybody needs everything; however, communication between both groups and systems (programs) is becoming more and more essential. My plea, therefore, is for all of us to take very seriously cooperative ventures in system building.

To this end, let me suggest that a first project area that needs attention is to show how an automated proof system can take advantage of a special-purpose "black box" for doing symbolic algebra. There has been very extensive development of algebra systems over the last years amounting to many man-years of work, and there is no reason for a proof system to engage in a bottom-up re-implementation of programs that are already available and working very efficiently. The problems of syntactic interfacing that are needed to get two such programs to communicate strike me as being easy to solve. In any case, the methods of logical proof are not the right methods to apply to problems of algebraic simplification. But proofs can be made much more practical if they can call on simplification procedures.

By the same token, it is very clear that a system such as *Mathematica* – or more precisely, a person who is trying to write *Mathematica* programs and use such a system for research – badly needs a proof system (even for elementary properties of integers, reals, or complexes) to get their programs formulated in the most compact manner. Also many small details have to be checked *by proof* even to get programs correct. I am not calling here for automatic program verification at all. What we badly need is a "programmer's mathematical assistant" for providing good interactive programming environments. We also need to make these environments usable by ordinary scientists who are not going to be trained computer hackers. More progress on this front ought to be given a high-priority for concerted effort by cooperating groups from all over the world – both in industry and at universities and research centres.